Oppositions and Paradoxes

Oppositions and Paradoxes

Philosophical Perplexities in Science and Mathematics

John L. Bell

broadview press

BROADVIEW PRESS— www.broadviewpress.com
Peterborough, Ontario, Canada

Founded in 1985, Broadview Press remains a wholly independent publishing house. Broadview's focus is on academic publishing; our titles are accessible to university and college students as well as scholars and general readers. With over 600 titles in print, Broadview has become a leading international publisher in the humanities, with world-wide distribution. Broadview is committed to environmentally responsible publishing and fair business practices.

The interior of this book is printed on 100% recycled paper.

Library and Archives Canada Cataloguing in Publication

Bell, J. L. (John Lane), author
Oppositions and paradoxes : philosophical perplexities in science and mathematics / John L. Bell.

Includes bibliographical references and index.
ISBN 978-1-55481-302-5 (paperback)

1. Paradoxes. 2. Opposition, Theory of. 3. Science—Philosophy. 4. Mathematics—Philosophy. I. Title.

BC199.P2B44 2016 165 C2016-900811-8

Broadview Press handles its own distribution in North America
PO Box 1243, Peterborough, Ontario K9J 7H5, Canada
555 Riverwalk Parkway, Tonawanda, NY 14150, USA
Tel: (705) 743-8990; Fax: (705) 743-8353
email: customerservice@broadviewpress.com

Distribution is handled by Eurospan Group in the UK, Europe, Central Asia, Middle East, Africa, India, Southeast Asia, Central America, South America, and the Caribbean. Distribution is handled by Footprint Books in Australia and New Zealand.

Broadview Press acknowledges the financial support of the Government of Canada through the Canada Book Fund for our publishing activities.

Edited by Robert M. Martin
Book design by Michel Vrana

PRINTED IN CANADA

To Sandra, who lights my way

Contents

Acknowledgements

It is a pleasure to thank Bryson Brown, one of the original reviewers of the book. His careful reading of the manuscript and his insightful comments on it have led to many refinements in my presentation. My thanks also to Bob Martin, the copy editor at Broadview Press, for his scrutiny of the text and his constructive suggestions for improving it.

My wife Sandra listened graciously while I read sections of the manuscript to her. I am most grateful for her trenchant comments, which have resulted in numerous clarifications and helped to straighten out the sometimes knotty style of my writing.

Finally, I would like to thank Stephen Latta, the Philosophy Editor at Broadview Press. His enthusiasm and efficient handling of the publication process has been an author's joy.

What Is This Book About?

At some point in the remote past human beings began to think in terms of general ideas, or *concepts*. It must have occurred to those dawning intellects that concepts could be in *opposition* to one another. Thus emerged the opposed concepts *The One and the Many*, *The Finite and the Infinite*, *The Bounded and the Unbounded*, *The Continuous and the Discrete*, *The Subjective and the Objective*, *The Absolute and the Relative*, *The Whole and the Part*, *The Constant and the Changing*, *The Straight and the Curved*, *Chance and Necessity*, *Being and Nothingness*. From antiquity, opposed concepts such as these have been a driving force in philosophical, scientific, and mathematical thought. Yet they have also given rise to perplexities and paradoxes which continue to haunt conceptual thinking to this day. This is what the book you are about to read is about.

We begin with *The Continuous and the Discrete*, an opposition that has underlain the development of mathematics and science since antiquity. Next, we turn our attention to *Set Theory*, the fundamental branch of mathematics which rests on the One/Many, Finite/Infinite, and Whole/Part oppositions. This is followed by a discussion of the strange universe of *Non-Euclidean Geometry*, the revolutionary form of geometry developed in the nineteenth century which embodies the Straight/Curved and Absolute/Relative oppositions. We then proceed to an analysis of the puzzles and paradoxes associated with the idea of *Time Travel*. As we show, time travel involves a number of oppositions, including One/Many, Subjective/Objective, and Chance/Necessity. Next, we discuss *Einstein's Theory of Relativity*, in which the Absolute/Relative, Constant/Changing, and Bounded/Unbounded oppositions may be seen. After this we come to *Quantum Physics* which again embodies a number of oppositions, in particular Continuous/Discrete, Chance/Necessity, Absolute/Relative, and Whole/Part. The main text concludes with a discussion of *Cosmology*, a subject in which many oppositions figure, in particular, Finite/Infinite, Whole/Part, One/Many, Constant/Changing, Chance/Necessity, Being/Nothingness. At the end of the book are to be found a number of Appendices, including discussions of logical and linguistic paradoxes, Constant/Changing oppositions, and oppositions in Kant's philosophy.

Understanding the bulk of the contents of this book requires an acquaintance with no more than elementary mathematics. Those few places where more advanced mathematical knowledge is desirable, for example, Appendix 4, can be skipped without serious loss.

Chapter I

The Continuous and the Discrete

CONTINUITY AND DISCRETENESS

We begin with a fundamental conceptual opposition: *the Continuous and the Discrete.*

Everyone is familiar with the idea of *continuity*. To be continuous is to form an unbroken or uninterrupted whole, like the ocean or the sky, or the act of breathing. The word "continuous" derives from a Latin root meaning "to hang together" or "to cohere." This same root gives us the nouns "continent"—an expanse of land unbroken by sea—and "continence"—self-restraint in the sense of "holding oneself together." Synonyms for "continuous" include: *connected, entire, unbroken, uninterrupted.*

A continuous entity—a *continuum*[1]—has no "gaps."

1 The term "continuum" has become a buzzword, especially popular with science fiction writers. This was surely the result of (continued) exposure to the phrase "space-time continuum" with which popular accounts of relativity theory are replete.

Opposed to continuity is *discreteness*: to be discrete is to be separated, like the scattered pebbles on a beach or the leaves on a tree. The word "discrete" derives from a Latin root meaning "to separate." This same root yields the verb "discern"—to recognize as distinct or separate—and the cognate "discreet"—to show discernment, hence "well-behaved." Synonyms for "discrete" include *separate, distinct, detached, disjunct.*

It is a curiosity of language that "continuity" and "discreteness" are antonyms, yet "continence" and "discreetness" are synonyms.

Continuity implies unity; discreteness, plurality.

While it is the fundamental nature of a continuum to be *whole* or *undivided*, it is generally held that any continuum can be repeatedly, even endlessly divided without altering its essential nature. So, for instance, the water in a bucket may be repeatedly halved and yet remain water. (For the purposes of argument we are here ignoring the atomic nature of matter established by modern physics.) Granted this, it follows that the process of dividing a continuum into ever smaller parts will never terminate in an *indivisible* or an *atom*—that is, a part which, lacking proper parts itself, cannot be further divided. That is, a continuum is *limitlessly* or *endlessly divisible.*

Discrete entities, on the other hand, typically cannot be divided without eventually effecting a change in their essential nature. For example, consider the pair of wheels on a bicycle. If we divide this pair into two, we obtain two separate wheels. If now we take one of these wheels and divide *it* into two, we obtain, not a wheel, but a *half-wheel.* This is plainly no longer a wheel. Unlike water, wheels are indivisible things.

Thus we are presented with two opposed properties: on the one hand, the property of being indivisible, separate or discrete; and, on the other, the property of being endlessly divisible and continuous although not actually divided into parts.

There is a sense in which a thing can be *both* continuous *and* discrete. For example, if a wheel is regarded simply as a piece of matter—a *mass*—it remains so on being divided in half. In other words, the wheel regarded as a wheel is discrete, but regarded as a mass, it is continuous. This shows that continuity and discreteness are in fact complementary properties originating through the mind's ability to perform acts of abstraction, that is, to ignore everything but essentials.

Continuity arises from abstracting a thing's endless divisibility, discreteness from abstracting its self-identity.

The opposition between discreteness and continuity is reflected in linguistics, where a noun such as "wheel" denoting a discrete thing is called a *count noun*, while a noun denoting a continuous thing such as "water" is called a *mass noun*. Count nouns, like discrete things, admit plurals. Mass nouns, on the other hand, ordinarily do not. Thus we say "two wheels" but not "two waters."[2]

In mathematics the idea of discreteness is embodied in *arithmetic*, that is, in the concept of *whole number*, or (positive) *integer*. Numbers are associated with the idea of a collection of separate individual objects, whose properties—*apart from their distinctness*—have been ignored. When we say "here are five apples," it is the distinctness of the individual apples, not the fact that they happen to be apples (or anything else), which legitimates the use of the number term "five." From the standpoint of mathematics, arithmetic is the realm of the discrete.

The primary mathematical embodiment of the idea of continuity, on the other hand, is found in *geometry*, where it takes the form of *geometric figure*. Because geometric figures such as straight lines, circles, and spheres are continuous, they can, in principle at least, be endlessly divided. The endless divisibility of geometric figures plays an indispensable role in the development of geometry in Euclid's *Elements*. From the standpoint of mathematics, geometry is the realm of the Continuous. The Discrete arises in geometry through the concept of a *point*, which Euclid defined as *that which has no part*. Points mark the boundaries of lines. While lines have length, points, as partless boundaries of lines, are lengthless.

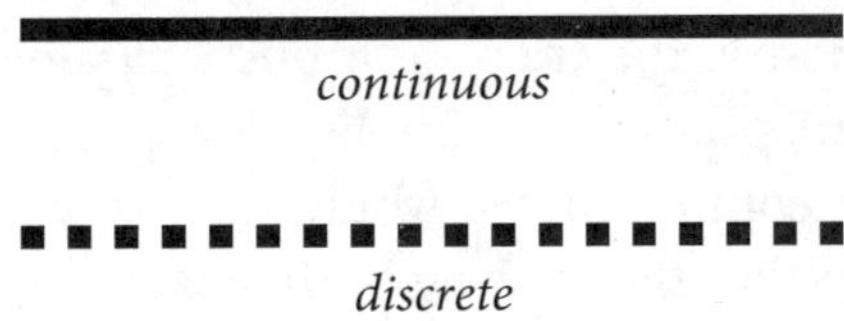

The dominion of the Discrete is a model of tidiness in which quality is reduced to quantity and over which the concepts of number and

2 In a restaurant one might ask the waiter for "two waters," but this is actually an abbreviated form of "two glasses or bottles of water," in which the count nouns "glass" or "bottle" figure.

point reign supreme. It is populated by point-like units, which, lacking inherent qualities, are intrinsically indistinguishable from one another. Difference is manifested through plurality alone: single points in themselves are identical, but a collection of two or more points is easily distinguished from a single point. The purity of the principles governing discreteness has led philosophers to regard the Discrete as a paragon of clarity, a domain within which pure reason, freed of the entanglements of qualities, can be realized to its fullest extent.

By contrast, the realm of the Continuous is a jungle. It teems with such exotic entities as incommensurable lines, horn angles, space curves, one-sided surfaces, and fractals. The taming of the Continuous by reduction to the Discrete has been a principal task, if not *the* principal task, of mathematics.

Yet the jungle of the continuous is grounded in such apparently simple notions as space, time, and extension, continua at the very core of our experience. And certain philosophers have maintained that all natural processes occur continuously: witness, for example, Leibniz's famous assertion *natura non facit saltus*—"nature makes no jump."

Since any continuum is endlessly divisible, dividing it successively will generate an unbounded number of parts. In antiquity this claim met with the following objection. If the process of dividing an extended magnitude, such as a continuous line, could be carried out *completely*, then the magnitude would be reduced to a multitude of lengthless points (or even, possibly, to nothing at all). No matter how many such points there may be—even if infinitely many—they cannot be "reassembled" to form the original magnitude, since a sum of lengthless elements still lacks length.[3]

This problem is brought into sharp focus when the magnitude undergoing division is the track of a body passing from one point to another in a finite time. If infinitely many points remain after the division has been completed, we would have to draw the seemingly absurd conclusion that the moving body passes over infinitely many points or locations in a finite time. This is one of *Zeno's paradoxes* which we shall soon take up.

---

3 Of course, this presupposes that there are no "gaps" between the elements or points, which is implicit in the assumption that the points have been obtained by complete division of a continuum.

THE PYTHAGOREAN SCHOOL AND INCOMMENSURABLE MAGNITUDES

Continuity and discreteness are united in the process of *measurement*. In measuring the lengths of lines, for example, a discrete unit such as an inch or a metre is chosen. But then it is found that most lines fail to have lengths which are exact integral multiples of the chosen unit, and so cannot be "counted" by that unit. The ancient Egyptians and Babylonians, seeing this, were led to invent the idea of a *fraction*, that is, a *part* of a unit. Parts of given units can serve as new units enabling counting by units to carry on. Thus the humble fraction is the product of two oppositions: the Continuous and the Discrete, and the Whole and the Part.

The introduction of fractions by the ancients amounted to an apparent overcoming of the opposition between the Continuous and the Discrete. But such peaceful coexistence was eventually to give way to a striking reaffirmation of the fundamental opposition between the two concepts. This erupted with the discovery of *incommensurable magnitudes* by the Pythagorean school in Greece around 550 BCE.

The Pythagoreans had made remarkable advances in mathematics. It is believed that they coined the term "mathematics" from a root meaning "learning" or "knowledge." Their mathematical achievements convinced them that mathematics, and more especially *number*,[4] is the very essence of the real. Pythagoreanism may justly be called the first attempt to describe the world in mathematical, that is, scientific, terms.

In the Pythagorean vision the structure of mathematics took the form of a bifurcating scheme of oppositions:

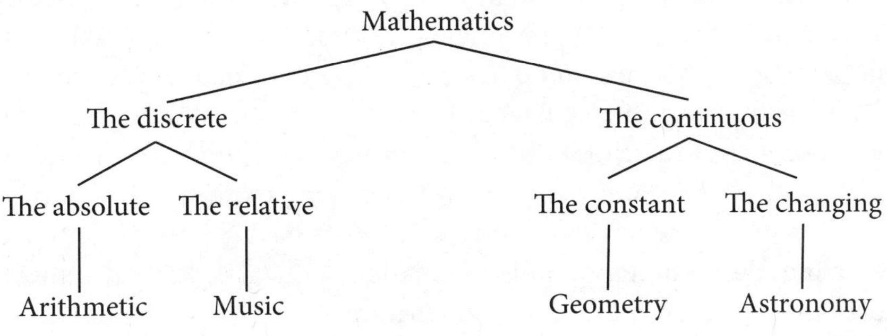

4 It should be mentioned that the Pythagoreans were also number mystics, ascribing magical properties to numbers.

The presence of music in the Pythagorean scheme as the "relative discrete" arises from their discovery of the correspondence between musical intervals and simple arithmetical ratios of lengths of stretched strings. In this correspondence the octave is associated with the ratio 2:1, the fifth with 3:2, and the fourth with 4:3. This discovery, which later led to the construction of musical modes and scales, is particularly striking in being both a contribution to *mathematical physics*, the mathematical description of the natural world, and to *aesthetics*, the analysis of the beautiful.

The bottom line of the Pythagorean scheme—arithmetic, music, geometry, and astronomy—constitutes the so-called *quadrivium*. This, conjoined with the *trivium* of grammar, logic, and rhetoric, made up the *seven liberal arts*, the basis of Western pedagogy up to the Middle Ages. Today only arithmetic and geometry are still regarded as branches of mathematics.

It was a fundamental principle of Pythagoreanism that all is explicable in terms of properties of, and relations between, whole numbers, that is in terms of the Discrete. In particular they upheld what may be termed the *principle of commensurability*. This is to the assertion that, in comparing the sizes of two similar magnitudes such as the lengths or areas or volumes or masses, an appropriate unit can always be chosen so that the ratio of the sizes of the two magnitudes can be expressed as a numerical ratio, that is, a fraction. It must have been a great shock to the Pythagoreans to find, as they did, that this principle *cannot be upheld within geometry itself*. This followed upon the shattering discovery, probably made by the later Pythagoreans before 410 BCE, that ratios of whole numbers *do not suffice* to allow the length of the diagonal of a square to be compared with the length of its side. They showed that these line segments are *incommensurable*, that is, it is impossible to choose a unit of length sufficiently small so as to enable both the diagonal of a square and its side to be measured by an integral number of the chosen units. The discovery of incommensurable lines dealt a devastating blow to the whole Pythagorean philosophy.

The Pythagorean catastrophe shows that basing a world outlook on a literal interpretation of mathematics can be risky. The exactness of mathematics may eventually undermine it!

The initial steps of what may have been the Pythagorean proof of the incommensurability of the side and diagonal of a square are presented in Plato's dialogue the *Meno*. Here Socrates asks an uneducated slave how to double the area of a square while maintaining its shape. Eventually the problem is solved by producing the figure below, in which the inscribed square *BDIH* has double the area of the given square *ABCD*.

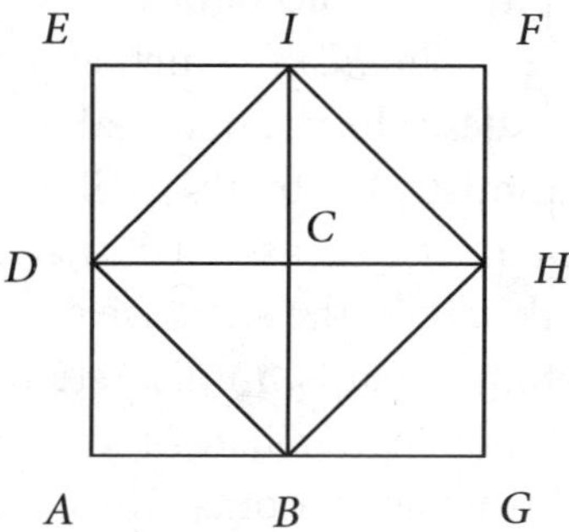

Having achieved his purpose, Socrates stops at this point, but it is possible that the Pythagoreans continued the argument along the following lines. Suppose *AB* and *BD* are commensurable; then we may take a unit of measure which divides both *AB* and *BD* an integral number of times, and, by successively doubling the length of the unit, if necessary, we may assume that at least one of these numbers is odd. Then the area of *AGFE* is divisible by 4. But *AGFE* = 2 *BHID*, so that the area of *BHID* is even. Now the area of *BHID* is BD^2, so the latter is even, and it follows that *BD* is even. Therefore *AB* must be odd. Now it can be seen from the figure that the area of *BHID* is divisible by 4. But *BHID* = 2 *ABCD* so the area of *ABCD* is even. But this area is AB^2, and so *AB* must be even. This is a contradiction, and we conclude that *AB* and *BD* are incommensurable.

In modern terms, the Pythagorean incommensurabilities are understood as asserting that the "numbers" representing the lengths of certain lines are *irrational*, that is, are not rational fractions of the form m/n with integral m, n. Thus, for example, the diagonal of a square of side 1 has length $\sqrt{2}$, and we may establish the irrationality of the latter by means of the following argument, which is essentially an algebraic version of the one given above. Suppose that $\sqrt{2}$ is rational. Then $\sqrt{2} = m/n$ with integral m, n. We may take m/n to be in lowest terms so that at least one of m, n is odd. Then $2 = m^2/n^2$ so that $m^2 = 2n^2$. Hence m^2 is even, and so therefore is m. It follows that n must be odd. But since m is even, m^2 is divisible by 4, so that $n^2 = m^2/2$ is even and so therefore is n. Thus n is both even and odd, a contradiction. It follows that $\sqrt{2}$ cannot be rational.

ATOMISM

The opposition between the Continuous and the Discrete also figures in the ancient Greek physicists' account of the nature of *matter*, or *substance*. The doctrine of *atomism*, which emerged in the fifth century BCE, purported to describe the material world entirely in terms of the Discrete. The atomists, led by the philosopher Leucippus and the mathematician Democritus, asserted that matter is *not* limitlessly divisible. Not only would the successive division of matter ultimately terminate in atoms, that is, in discrete particles incapable of being further divided, but matter had *in actuality* to be conceived as being compounded from such atoms. The atomists held that material form results from the arrangement of the atoms constituting all matter. They further held that the sole form of motion is the motion of individual atoms in a vacuum, and that physical change can only occur through the mutual impact of atoms.

In attacking limitless divisibility the atomists were at the same time mounting a claim that the continuous is ultimately reducible to the discrete, whether at the physical, theoretical, or perceptual level.

Atomism was later to flower into a general doctrine of the reducibility of the complex to the simple. In this regard there are a number of versions of atomism in addition to the physical atomism of the ancients. These include

- *Mathematical* atomism, the doctrine—originating with the Pythagoreans of the sixth century BCE—that all mathematical concepts are ultimately reducible to numbers

- *Logical* atomism, the positing of atomic or elementary propositions from which all meaningful assertions can be built up

- *Biological* atomism, the postulation of discrete organic units such as cells or genes determining the structure of all organisms

- *Linguistic* atomism, the principle of constructing words from the letters of an alphabet and sentences from the words produced thereby[5]

---

5 Joseph Needham has suggested that the emergence of atomism is connected with the alphabetic principle on which the great majority of natural (written) languages

THE STOICS AND THE CONTINUUM THEORY OF MATTER

Two centuries after the birth of the discrete doctrine of atomism, a rival physical theory based on the Continuous—the *continuum theory*—was developed by the Stoic school of philosophy led by Zeno of Cition and Chrysippus.

The Stoics postulated that the ultimate foundation for all natural phenomena is a limitlessly divisible continuous substance pervading all space they called *pneuma* (Greek for "breath"). This substance was regarded as a kind of amalgam of air and fire, two of the four elements (the others being earth and water) which the fifth-century BCE philosopher Empedocles had proposed as the fundamental constituents of matter. Pneuma was conceived as being an elastic medium through which impulses are transmitted by wave motion.[6] All physical occurrences were taken to be linked through tensile forces in the pneuma, and matter itself was held to derive its qualities from the "binding" properties of its indwelling pneuma.

A major problem encountered by the Stoic philosophers was that of the nature of *mixtures*. They struggled to explain how the pneuma mixes with material substances so as to "bind them together." The atomists' granular conception of matter enabled them to avoid this difficulty, since they could regard a mixture of two substances as being just a combination of their constituent atoms into a complex—a kind of lattice or mosaic. But the idea of a mixture posed difficulties for the Stoic conception of matter as continuous. For in order to mix fully

---

rest. In his great work, *Science and Civilization in China* the parallel is noted between the limitless variety of words formable from the relatively few letters of the alphabet, and the idea that a very small number of "elementary" particles could, in a multitude of combinations, engender the limitless variety of material bodies. But in China atomism never really took root. In this connection Needham observes, "the Chinese written character is an organic whole, a Gestalt, and minds accustomed to an ideographic language would perhaps hardly have been so open to the idea of an atomic constitution of matter." As Needham points out, however, the Chinese recognized the function of the atomic principle in numerous contexts, for example the reduction of written characters to radicals, the composition of melodies from the notes of the pentatonic scale, and the representation of Nature through the permutations and combinations of the broken and unbroken lines in the hexagrams of their ancient work of divination the *I Ching*.

6 This aspect of the Stoic physics is a remarkable anticipation of the ether theory developed by physicists in the nineteenth century.

two continuous substances, the substances would either have to interpenetrate in some mysterious way, or, failing that, they would each need to be subjected to an infinite division into infinitesimally small elements which would then have to be arranged, like finite atoms, into some kind of discrete pattern.

This controversy over the nature of mixture shows that the idea of continuity is inextricably entangled with the puzzles of limitless divisibility and of the infinitesimally small. The mixing of particles of finite size, no matter how small they may be, presents no fundamental difficulty. But this is no longer the case when dealing with a continuum, whose parts can be divided *ad infinitum*.

ZENO'S PARADOXES

The problem of the limitless divisibility of continua had already been posed in a dramatic, yet subtle way more than a century before the rise of the Stoic school, by Zeno of Elea, a pupil of Parmenides, who taught that the universe was a static unchanging unity. Zeno's arguments take the form of *paradoxes* which are collectively designed to undermine the belief in motion, and hence in any kind of change. Zeno's paradoxes embody not only the opposition between the Continuous and the Discrete, but also that between the Finite and the Infinite.

The first two of Zeno's paradoxes, both of which rest on the assumption that space and time are continuous, are designed to prove that motion is impossible. This is done by showing that a finite motion generates an actual infinity.

The first paradox, the *Dichotomy*, goes as follows. Consider a body moving from one position to another. It is clear that, before the body can reach a given point in its path, it must first traverse half of the distance to that point. But before it can traverse half of that distance, it must traverse one quarter; and so on *ad infinitum*. So, for a body to pass from one point *A* to another, *B*, it must traverse an infinite number of divisions. But an infinite number of divisions cannot be traversed in a finite time, and so the goal cannot be reached.

The second paradox, *Achilles and the Tortoise*, is the best known. Achilles and a tortoise run a race, with the latter enjoying a head start. Zeno asserts that no matter how fleet of foot Achilles may be, he will never overtake the tortoise. For, while Achilles traverses the distance from his starting-point to that of the tortoise, the tortoise advances a certain distance, and while Achilles traverses this distance, the tortoise makes a further advance, and so on *ad infinitum*. Consequently Achilles will run forever without overtaking the tortoise.

This second paradox is formulated in terms of two bodies, but it has a variant involving, like the *Dichotomy*, just one. To reach a given point, a body in motion must first traverse half of the distance, then half of what remains, half of this latter, and so on *ad infinitum*, and again the goal can never be reached. This version of the *Achilles* exhibits a pleasing symmetry with the *Dichotomy*. For the former purports to show that a motion, once started, can never stop; the latter, that a motion, once stopped, can never have started.

The conclusion of the *Dichotomy* that motion can never get started is ingeniously illustrated by a modern variant of the paradox—the *paradox of the gods*, devised by the philosopher José Bernardete. Here someone attempts to walk a mile from *A* to *B*. A god waits in readiness to raise a barrier blocking the walker's further advance when she reaches the half-mile point. A second god (unknown to the first) waits in readiness to raise a barrier of her own when the walker has reached the quarter-mile point. This procedure is repeated by a third god, etc. *ad infinitum*. So the walker cannot even get started, because however short a distance she travels she will already have been stopped by a barrier. But in that case *no* barrier will rise, so that there is nothing to stop her from setting off. She has been forced to stay where she is by the mere unfulfilled intentions of the gods! Bernardete suggests that the strangeness of this conclusion can be mitigated by replacing the gods by laws of nature. The intention of the first god becomes the first law of nature that a barrier will appear at the half-mile mark, the second god's intention the second law of nature that a barrier will appear at the quarter-mile mark, etc. In that case the combined intentions of the gods is manifested as a field of force preventing the walker from moving.

The third of Zeno's paradoxes, the *Arrow*, rests on the assumption of the *discreteness of time*. Here we consider an arrow flying through the air. Since time has been assumed discrete we may "freeze" the arrow's motion—as in a film frame—at an indivisible instant of time. For it to move during this instant, time would have to pass, but this would mean that the instant contains still smaller units of time, contradicting the indivisibility of the instant. So at this instant of time the arrow is at rest. Since the instant chosen was arbitrary, the arrow is at rest at any instant. In other words, it is always at rest, and so motion does not occur.

Modern mathematics offers solutions to these paradoxes. In the case of the *Dichotomy*, the presentation may be simplified by assuming that the body is to traverse a unit spatial interval—a mile, say—in unit time—a minute, say. To accomplish this, the body must first traverse half the interval in half the time, before this one-quarter of the interval in one-quarter of the time, etc. In general, for every subinterval of length $1/2^n$ ($n = 1, 2, 3, \ldots$), the body must first traverse half thereof, i.e., the subinterval of length $1/2^{n+1}$. In that case both the total distance traversed by the body and the time taken is given by the convergent series

$$\sum_{n=0}^{\infty}\left(\frac{1}{2^n}-\frac{1}{2^{n+1}}\right)=\sum_{n=0}^{\infty}\frac{1}{2^{n+1}}$$

which sums to 1 as expected. So, *contra* Zeno, the infinite number of divisions is indeed traversed in a finite time.

More troubling, however, is the fact that these divisions, of lengths 1/2, 1/4, 1/8,... (in reverse order) constitute an infinite *regression* which, like the negative integers, *has no first term*. Zeno seems to be inviting us to draw the conclusion that it cannot be supplied with one, so that the motion could never get started. However, from a strictly mathematical standpoint, there is nothing to prevent us from placing 0 before all the members of this sequence, just as it could be placed, in principle at least, before all the negative integers. Then the sequence of correlations (1/2, 1/2), (1/4, 1/4),(1/8, 1/8),... (in reverse order) between the time and the body's position is simply preceded by the correlation (0, 0), where the motion begins. There is no contradiction here.

In the case of the *Achilles*, let us suppose that the tortoise has a start of 1000 feet and that Achilles runs ten times as quickly. Then Achilles must traverse an infinite number of distances—1000 feet, 100 feet, 10 feet, etc.—and the tortoise likewise must traverse an infinite number of distances—100 feet, 10 feet, 1 foot, etc.—before they reach the same point simultaneously. The distance of this point in feet from the starting points of the two contestants is given, in the case of Achilles, by the convergent series

$$\sum_{n=0}^{\infty} 1000\cdot 10^{-n}$$

which sums to 111 1/9 feet. In the case of the tortoise the corresponding distance is given by the convergent series

$$\sum_{n=0}^{\infty} 100 \cdot 10^{-n}$$

summing to 111 1/9 feet. And, assuming that Achilles runs 10 feet per second, the time taken for him to overtake the tortoise is given, in seconds, by the convergent series

$$\sum_{n=0}^{\infty} 100 \cdot 10^{-n}$$

summing to 111 1/9 seconds. Thus, again *contra* Zeno, Achilles overtakes the tortoise in a finite time.

Although the use of convergent series does confirm what we take to be the evident fact that Achilles will, in the end, overtake the tortoise, a nagging issue remains (pointed out by Bertrand Russell in his entertaining essay "Mathematics and the Metaphysicians"[7]). For consider the fact that, at each moment of the race, the tortoise is somewhere, and, equally, Achilles is somewhere, and neither is ever twice in the same place. This means that there is a one-one correlation between the locations of the tortoise and those of Achilles, so that the number of the tortoise's locations and those of Achilles must be the same. But when Achilles catches up with the tortoise, the positions occupied by the latter are only *part* of those occupied by Achilles. This would be a contradiction if one were to insist, as did Euclid in his *Elements*, that the whole invariably has more terms than any of its parts. In fact, it is precisely this principle which, in the nineteenth century, came to be repudiated for infinite sets such as the ones encountered in Zeno's paradoxes. This issue will be discussed in the next chapter. Once this principle is abandoned, no contradiction remains.

The paradox of the *Arrow* can be resolved by developing a theory of *velocity*, based on the differential calculus, in turn based on the theory of limits. By definition, (average) velocity is the ratio of distance travelled to time taken. In this definition *two* distinct points in space and *two* distinct points in time are required. Velocity at a point

7 In *Mysticism and Logic* (London: Pelican Books, 1954).

is then defined as the *limit* of the average velocity over smaller and smaller spatiotemporal intervals around the point. According to this definition, a body may have a nonzero "velocity" at each point, but at each instant of time will not "appear to be moving."

While Zeno's paradoxes can be resolved from a strictly *mathematical* standpoint, they present difficulties for understanding the nature of *actual* motion which have persisted to the present day.

CONTEMPORARY VERSIONS OF ZENO'S PARADOXES: SUPERTASKS

A number of contemporary versions of Zeno's paradoxes involve the idea of a *supertask*, a notion (and term) introduced in 1954 by the English philosopher James Thomson. A supertask is an infinite sequence of operations O_1, O_2, ..., O_n, ... carried out in a finite time. Zeno's *Achilles* paradox may be construed as the carrying out of a supertask. It is easiest to see this in its alternative formulation, where a body in motion must first traverse half of the distance, then half of what remains, half of this latter, and so on *ad infinitum*. In that case, the operation O_n may be taken to be the movement of the body over the nth segment of its motion, of length $1/2^n$.

The question is: could a supertask possibly be carried out?

In *Thomson's lamp paradox*, we are given a lamp with an on-off switch. Flicking the switch once turns the lamp on. Another flick turns it off. Now assume the presence of a demon capable of carrying out the following task. The lamp is initially off. After half a minute has elapsed, it turns the lamp on. After another quarter minute, it turns the lamp off again. At the end of a further eighth of a minute, it turns the lamp on. The demon continues in this manner, flicking the switch each time after waiting exactly one-half the time it waited before the previous flick of the switch. The sum of this infinite series of time intervals is exactly one minute.

Is the lamp switch on or off after exactly one minute has elapsed?

Thomson himself thought that attempting to answer this question led to a logical contradiction, and so concluded that it is logically impossible to carry out a supertask. He argued that, at the end of one minute, the lamp is either on or off. If the lamp is on, then there must have been some last time, just before the one-minute mark, at which it was flicked on. But, such an action must have been followed by a flicking off action since, after all, every action of flicking the lamp on before the one-minute mark is followed by one at which it is flicked off between that time and the one-minute mark. So the lamp cannot be on and so must be off. Analogously, one can also reason that the lamp cannot be off, and so must be on at the one-minute mark. Therefore the lamp must be both on and off at the end of one minute. This is a contradiction. Thus it is logically impossible to carry out this supertask—and so any supertask.

As later commentators have observed, this argument is flawed. For while the lamp's state (on or off) is specified for every instant *strictly* before one minute has elapsed, the lamp's state *at the one minute mark* is unspecified, and is in fact indeterminate: at the end of the minute, the lamp could be either on or off. This shows that the completion of this supertask is at least logically possible.

Let **0** (**1**) represent the lamp being off (on), write [0, 1) for the half-open interval of instants strictly before one minute has elapsed, and [0, 1] for the closed interval of instants including the minute mark. Then the scenario can be represented by the function f defined on [0, 1) by

$$f(t) = \begin{cases} 0 \text{ if } t \text{ is in a subinterval of the form } [1—1/2^{2n}, 1—1/2^{2n+1}) \ (n = 0, 1, 2, ...) \\ 1 \text{ if } t \text{ is in a subinterval of the form } [1—1/2^{2n+1}, 1—1/2^{2n+2}) \ (n = 0, 1, 2, ...) \end{cases}$$

This function f is obviously discontinuous, and can be extended to [0, 1] arbitrarily by taking $f(1) = 0$ or $f(1) = 1$. The lamp's state at $t = 1$ is in no way determined by its previous states.

On the other hand suppose we arrange to make the lamp's state continuous by installing a rotating knob controlling the lamp's brightness, so enabling it to vary continuously from off (0) to full brightness (1). Then the lamp's range of brightness is represented by the closed interval [0, 1]. Now consider the function g: $[0, 1) \to [0, 1]$ defined by $g(t) = t$. Clearly g is continuous. If we think of $g(t)$ as representing the lamp's brightness at time t, then at $t = 0$, the lamp is off, at $t = 1/2$, the lamp is 50% bright, at $t = 3/4$, it is 75% bright, etc. Now, g can be arbitrarily extended to [0, 1] by taking $g(1)$ to be *any* value in [0, 1]. But the only *continuous* extension of g to [0, 1] is obtained by taking $g(1) = 1$. So if we insist that the brightness of the lamp vary continuously, the state of the lamp at the end of one minute—fully on—is uniquely determined by its states at the previous instant. This continuous version of Thomson's lamp isn't really a supertask, of course, because it can be (approximately, at least) realized in practice—just as, in the world of experience, one can move continuously between two points.

While the completion of supertasks may be *logically* possible, it may still be asked whether their completion is *physically* possible. General relativity, through the curious physics of black holes, seems to offer such possibilities, in principle at least. We shall defer discussing this matter until Chapter V.

INFINITESIMALS

In the seventeenth century the opposition between the Continuous and the Discrete erupted in mathematics once again, this time in the form of the *differential and integral calculus*. Up to the beginning of the twentieth century the calculus was known as the *infinitesimal calculus*. In the 11th edition of the *Encyclopedia Britannica* published in 1913, we find the following definition:

> The infinitesimal calculus is the body of rules and processes by which continuously varying magnitudes are dealt with in mathematical analysis. The name "infinitesimal" has been applied to the calculus because most of the leading results were first obtained by means of arguments about "infinitely small" quantities; the "infinitely small" or "infinitesimal" quantities were vaguely conceived as being neither zero nor finite but in some intermediate, nascent or evanescent state.

The central problem of the differential calculus—which actually originates in the opposition between the Constant and the Changing—is to determine the *rate of change* of a varying quantity, such as the location of a moving body varying with time. This rate of change is the *velocity* of the body. Now there is no difficulty in defining the *average velocity* of a moving body over an interval of time: one simply divides the distance travelled by the body by the length of the time interval, so that the average velocity is the ratio of distance travelled to time taken. But what can be meant by the velocity of a body at an *instant* of time: its *instantaneous* velocity? If the body is moving at a constant speed, then the instantaneous velocity at any instant can simply be identified with the unchanging average velocity. But what if the body's motion is *not* uniform? What can then be meant by instantaneous velocity?

The answer to this question dreamed up by seventeenth-century mathematicians was to define instantaneous velocity to be average velocity over an extremely short—an *infinitesimal*—interval of time. During an infinitesimal interval of time, a body will have moved an infinitesimal distance, and the instantaneous velocity is the ratio of the infinitesimal distance travelled to the infinitesimal time taken. In general, if the value y of one quantity varies systematically with the value of another quantity x, *the instantaneous rate of change* of y with respect to x is defined as the ratio of the infinitesimal change in y corresponding to an infinitesimal change in x.

A pretty solution to the problem. But the mathematicians of the seventeenth and eighteenth centuries were then confronted with the question: what, precisely, *is* an infinitesimal? And does it make sense to talk about the ratio of one infinitesimal to another? This question was to vex mathematicians for almost three centuries.

The mathematicians came up with several distinct, but closely related conceptions of the infinitesimal. One such conception—the *geometric* infinitesimal—may be considered as an embodiment of the opposition between the Continuous and the Discrete, since it results from the old idea of measuring continua in terms of discrete units. Geometric infinitesimals took two forms. The first comes from the insight that an infinitesimally small part of a curved line is indistinguishable from a correspondingly small part of a straight line, and so can actually be *treated as straight*. (Here we are presented with an instance of the opposition between the Straight and the Curved. More precisely, the claim is that, on the infinitesimal scale, the opposition between the Straight and the Curved no longer exists.) Thus, just as the perimeter of a polygon is the sum of its finite collection of edges, so any continuous curve can be considered as the "sum" of an infinite discrete collection of infinitesimally short linear segments—the "linear infinitesimals" of the curve. In particular, each closed curve can be thought of as an "infinilateral polygon," that is, a polygon with an infinite number of infinitesimal straight edges.

The second form of geometric infinitesimal arises analogously from the idea that a continuous surface or volume can be conceived as the sum of an indefinitely large, but discrete assemblage of lines or planes, the so-called *indivisibles* of the surface or volume. This idea, exploited systematically by the Italian mathematician Cavalieri in the seventeenth century, also appears in Archimedes' *Method* written during the third century BCE.

The other principal type of infinitesimal considered by seventeenth- and eighteenth-century mathematicians is the *arithmetical infinitesimal*, or *infinitesimal number*. As we have seen from the *Britannica* quotation above, this was conceived as a number which, while not zero, is in some sense smaller than any finite number. Mathematicians sometimes thought of an arithmetical infinitesimal as being represented by the "number" $1/\infty$, where ∞ is a symbol for infinity. The great seventeenth-century mathematician-philosopher G.W. Leibniz thought of arithmetic infinitesimals as *differentials*, infinitesimally small differences between finite numbers.

Still another notion of arithmetic infinitesimal arises in "practical" approaches to the differential calculus favoured by physicists and engineers. Here an infinitesimal is considered to be a quantity or number so small that its square and higher powers can be neglected, i.e., set to zero. These are known as *nilpotent* infinitesimals. The idea is discussed further in Appendix 4.

Despite their vague and puzzling nature, infinitesimals enjoyed a considerable vogue among seventeenth- and eighteenth-century mathematicians. As the charmingly named "linelets" and "timelets," they played an essential role in *Isaac Barrow's* (who was Newton's teacher) "method for finding tangents by calculation," which appears in his *Lectiones Geometricae* of 1670. As "evanescent quantities" they were instrumental (although later abandoned) in Newton's development of the calculus, and, as "inassignable quantities," in Leibniz's. The Marquis de l'Hôpital, who in 1696 published the first treatise on the differential calculus (entitled *Analyse des Infiniments Petits pour l'Intelligence des Lignes Courbes*), invokes the concept in postulating that "a curved line may be regarded as being made up of infinitely small straight line segments," and that "one can take as equal two quantities differing by an infinitely small quantity."

However, the conception of infinitesimals as genuine numbers was somewhat nebulous and even led, on occasion, to logical inconsistency. Memorably derided by the eighteenth-century philosopher George Berkeley as "ghosts of departed quantities" and later condemned by Bertrand Russell as "unnecessary, erroneous, and self-contradictory," this conception of infinitesimal gave way to the idea—originally suggested by Newton—of the infinitesimal as a *continuous variable* which becomes arbitrarily small. By the start of the nineteenth century, most mathematicians had come to accept this revised conception of infinitesimal. A line, for instance, was seen as consisting not of a collection of "points" but as the domain of values of a continuous variable, in which separate points are to be considered as locations. At this stage, then, the Discrete had been replaced by the Continuous.

But the development of mathematical analysis in the later part of the nineteenth century led mathematicians to demand still greater precision in the idea of a continuous variable, and above all in fixing the concept of *real number* as the value of an arbitrary such variable. As a result, a theory was developed in which a line is represented as a set of points, and the domain of values of a continuous variable by a set of real numbers. Within this scheme of things there was no place for the concept of infinitesimal, which accordingly left the scene for a time. Thus the Continuous was reduced to the Discrete and the properties of a continuum derived from the structure of its underlying collection of points. This reduction, underpinned by the development of set theory, met with almost universal acceptance by mathematicians.

This state of affairs remained in place until the 1960s, when infinitesimals again came to life. The logician Abraham Robinson, using methods of mathematical logic, created *nonstandard analysis*, a consistent theory of infinitesimals as infinitely small numbers, in something like Leibniz's sense, satisfying the usual laws of arithmetic. And the 1970s saw the development of *smooth infinitesimal analysis*, a consistent framework for mathematics based on geometric and nilpotent infinitesimals. Smooth infinitesimal analysis is the mathematics of a *perfectly smooth world*, in which all curves are smooth, that is, continuous and lacking jagged edges. This provides a modern, rigorous realization of the seventeenth-century idea that every curve is "locally straight" and that every closed curve can be thought as an "infinilateral polygon." Smooth infinitesimal analysis may be seen as marking a return in the foundations of mathematics from the Discrete to the Continuous.

Chapter II

Oppositions and Paradoxes in Mathematics: Set Theory and the Infinite

SET THEORY AND THE ONE/MANY OPPOSITION

The concept of the *infinite* is implicit in mathematics. The idea of infinity dawns in a child's mind when he or she asks whether there is a largest number. The child quickly grasps that, in principle, for every number which can be thought of, a larger one can be produced by the addition of 1. The sequence of numbers is seen to be *inexhaustible*, and so, in some sense, *unbounded*, that is, *infinite*. This is the source of the common idea of the infinite sequence of positive integers 1, 2, 3, ... Similarly, in geometry, the points, or straight lines, or circles in the Euclidean plane are also inexhaustible, and so infinite. In analysis, the collections of real or complex numbers are both infinite.

It was not until the nineteenth century that the mathematical infinite became the subject of precise analysis. The first steps in this direction were taken by the Austrian philosopher Bernard Bolzano in the first half of the nineteenth century, but his work went largely unnoticed at that time. The modern theory of the mathematical infinite—*set theory*—was created by the great German mathematician Georg Cantor in the latter half of the nineteenth century. Set theory has come to penetrate, and influence decisively, virtually every area of mathematics. It also plays a central role in the logical and philosophical foundations of mathematics.

As a theory of the mathematical infinite, set theory naturally embodies the opposition between the Finite and the Infinite. But set theory rests on a still more basic opposition, that between the *One and the Many*.

The *One and the Many* is the primary philosophical opposition. It is built into language as the distinction between *singular* and *plural*. It is the very source of Western philosophy, which begins with the question: is Being ultimately a unity or a plurality, a One or a Many? This question obsessed the earliest Greek philosophers. Around 500 BCE Heraclitus asserted that the One is made up of all things, and all things issue from the One. Somewhat later Parmenides declared that Being constitutes an indivisible, unchanging One. In the fourth century BCE Plato maintained that each set of many things to which we assign the same name is united by a single Form. The doctrine that all Being is fundamentally One is called *monism*. Its opposite, *pluralism*, is the doctrine that Being is, ultimately, Many.

Cantor described a set as "a many which allows itself to be thought of as a one." In 1895 he defined the idea of a set in the following way:

> By a set we understand every collection to a whole of definite, well-differentiated objects of our intuition or thought.

The fundamental question in Cantor's set theory is thus: *when can a collection be treated as a whole*, that is, as an individual thing? Let us give this question some thought.

Everyone is familiar with the process of *collecting* or *assembling* a number of objects into a single whole. For example, a dozen eggs are collected, arranged in a carton, and sold as a single entity. Each egg may be identified as a member of the carton (thought of not as the container, but rather as the collection of a dozen eggs). The singular term *carton* is used as a name for the plural term *a dozen eggs*. Thus we speak of a carton of eggs as an individual thing. Similarly, a nation's army is formed by assembling a number of citizens from that nation; it is then, for military purposes, treated as an individual thing.

From such examples it might be inferred that *any* collection of many individual things can *always* be treated as if it were an individual thing. But this would be a mistake, as *Russell's Paradox*—formulated by the philosopher Bertrand Russell in 1901—shows.

Let us follow Cantor in calling a *set* any collection of things which can be treated as an individual thing. Considered as an individual thing, a set can be a member of a collection. An individual carton of eggs in a grocery store, for example, is a member of the collection of all the cartons of eggs in the store, in just the same sense that each egg in the given carton is a member of that carton. Now in principle it is possible, in just the same sense, for a set to be a member of *itself*. Consider, for instance, the collection of all things which are not cats. If this collection could be treated as an individual thing, that is, if it were a set, then clearly—since it is not a cat—it would be a member of itself. However, "typical" sets, such as cartons of eggs, or armies, are *not* members of themselves—a carton of eggs cannot be a member of itself since the carton is not an egg, and an army cannot be a member of itself since the army is not a soldier.

Now we can formulate Russell's paradox. Call a set *typical* if it is *not* a member of itself. So a set is non-typical just when it *is* a member of itself. Now consider the collection **R**—*Russell's Collection*—of all typical sets. A set is a member of **R**, then, if and only if it is typical. Can **R** itself be a set? That is, can it be treated as an individual thing? The answer is no. For suppose that **R** were a set. Then **R** is either typical or non-typical. If **R** is typical then it is a member of itself, and so it is non-typical. If **R** is non-typical then it cannot be a member of **R**, and so is typical. Thus the assumption that **R** is a set leads to a contradiction. This is Russell's Paradox. We conclude that **R** cannot be treated as an individual thing.

We might well ask if the collection **S** of all *non-typical* sets can be treated as an individual thing, that is, can it be a set? (A set is non-typical if it *is* a member of itself.) If we assume that **S** is a set, and apply to it the same argument as for **R**, we can conclude only that, if **S** is typical, then it isn't a member of itself, and if **S** is non-typical, then it is a member of itself. This harmless conclusion is nothing more than the definition of "typical," and is certainly not self-contradictory. In fact it is perfectly consistent to suppose that **S** is a set. It can be either typical or non-typical.

A nice variation on Russell's paradox can be based on the idea of a *bibliography*, a list of titles of books. Bibliographies are typically supplied as appendices to works as lists of titles of (other) works which are referred to in the text. Now occasionally a work may be a list of titles of books, for example, catalogues of rare books offered at an auction. Such are examples of *pure* bibliographies, that is, books consisting of lists of book titles. It is then meaningful to ask whether a given (pure) bibliography *lists itself or not*. It now makes sense to consider the *bibliography of all bibliographies which don't list themselves*. At first sight such a bibliography would be self-contradictory, and so can't exist. For if it listed itself, then it would, in accordance with its own definition, not list itself, and *vice-versa*. This is the traditional "paradoxical" conclusion. But a more searching analysis yields a different perspective on the matter. Unlike libraries, bibliographies are not literally collections of books, but collections, or lists, of *titles* of books. So in order to analyze the paradox properly we need to suppose that we are given two domains of things: *titles (of books)* and lists of titles, or *bibliographies*. We shall suppose that each bibliography has been assigned a (unique) title. Now we can consider the bibliography **B** which lists just titles of bibliographies which don't list their own titles. Then we can see that **B** lists its own title; for, if not, its title, call it *b*, has the property that there is a bibliography, namely **B** itself, of which *b* is the title, and which doesn't list *b*. But, since **B** lists all such titles, it must therefore list *b*. We conclude that **B** lists its own title after all. Now since **B** lists its own title *b*, it follows from the definition of **B** that *b* must also be the title of a bibliography **C** which *doesn't list its own title*. Then, **B** and **C** have the same title (namely, *b*) but they *cannot* be identical, since **B** lists its own title but **C** does not. We conclude therefore that *there must exist two different bibliographies which are assigned the same title*. One of these, as we have seen, is **B**. **B** is a perfectly well-defined bibliography, and doesn't generate a paradox, *but it must necessarily have the same title as a different bibliography*. In other words, the assignment of titles to bibliographies must be *ambiguous* in the sense that some title is shared by two different bibliographies.

Note the difference between the conclusion here and that of Russell's paradox. In the latter case the conclusion is that there is a plurality which cannot be treated as an individual thing. In our analysis of the bibliographic case the conclusion is that, while each bibliography, considered as a plurality of titles, can be assigned a title—that is, in a certain sense treated as an individual—*there will always exist two distinct pluralities which have to be treated as if they were the same thing.* In both cases there are more pluralities than there are individuals.

We can also consider the bibliography—call it **A**—which lists just titles of bibliographies which *do* list their own titles. Can anything be inferred about **A** and its title *a* along the same lines as **B**? No, because the assumption that **A** doesn't list *a* merely leads to the stronger conclusion that there can be *no* bibliography with title *a* that lists *a*. And from the assumption that **A** does list *a* we can only infer the weaker statement that there exists a bibliography with title *a* that lists *a*. In fact nothing can be inferred about **A** and its title: **A** may or may not list its title; and it may or may not share a title with a different bibliography.

Another variant on Russell's paradox, suggested by Russell himself, is the *paradox of the barber*. In a certain town there lives an enterprising barber who resolves to shave just those men in the town who don't shave themselves. But then he asks himself: should I shave myself or not? He soon realizes that, in order to satisfy his own conditions, if he does shave himself, he mustn't, and if he doesn't shave himself, he should. He sees that, on pain of contradiction, he cannot satisfy his own requirements. The ambitious barber is thus, on logical grounds, forced to weaken his requirements and to content himself with shaving just those men in the town—*apart from himself*—who don't shave themselves. Whether the barber decides to shave himself or not is then entirely up to him—either choice is consistent.

A non-paradoxical, but amusing, variant on Russell's paradox arose in my university department recently. Sitting on top of the department mailboxes is a box labelled *mail for department members with no mailbox*. It can usually be seen to be overflowing with mail. Calling this box **P** (cleverly named the *parabox* by my colleague Anthony Skelton), **P** purports to store just mail for those department members who have no mailbox. Now if **P** were truly living up to its definition, then no department member could actually have mail stored there. For suppose that a department member, *M* say, had mail in **P**. Then there exists a mailbox, namely **P**, in which *M* has mail. In particular, *M* has a mailbox. But since **P** contains mail only for those *lacking* a mailbox, **P** cannot store *M*'s mail. This is a contradiction, and we conclude that *M* cannot have mail in **P**.

Accordingly no department member can have mail in **P**. It follows that every department member has a mailbox, for if she doesn't have a mailbox, then her mail would be stored in **P**; and as we have seen, that is impossible. But then **P** would have to be empty. This conclusion may seem strange, but actually it makes sense, since if everybody had a mailbox, then as a joke one could simply display a necessarily empty box labelled "mailbox intended precisely for those with no mailbox." This would be perfectly consistent. So **P** isn't "paradoxical" after all; its existence simply implies that it must necessarily be empty.

But the "parabox" is in fact *full* of mail. To resolve the apparent "parabox paradox," all we need to do is define a "regular" mailbox to be one with a department member's name on it and then label the "parabox" as "mailbox precisely for those with no regular mailbox." Then the mail actually in the "parabox" is—as intended—addressed only to those lacking a regular mailbox.

Russell's paradox shows that the basic question of Cantor's set theory—*when can a collection be treated as an individual thing?*—is much deeper than it appears at first glance.

Some light can be shed on this question by considering the process of *counting*. Here the members, or *elements*, of a given collection are counted one by one. When the process terminates, we obtain a positive integer called the *cardinal number* of the collection. For example, by counting the members of a carton of a dozen eggs, we determine that its cardinal number is 12. Each positive integer 1, 2, 3, ... is an *individual* thing associated (except in the case of 1) with a plurality.

Now notice that in determining the cardinal number of a collection we have assumed that the process of counting its members *terminates*. If this is the case, the collection is called *finite*. The process of counting the members of a finite collection, so determining its cardinal number, shows that the collection can be treated as a unity in *the numerical sense*. But clearly there also exist collections for which the process of counting their members *never terminates*, and so are *infinite*. Examples of infinite collections, such as the collection of positive integers 1, 2, 3, ... have been mentioned at the beginning of this chapter.

Since the process of counting an infinite collection never terminates, it cannot be assigned an *integer* as a cardinal number. Can *any* sense be made of the idea of assigning a cardinal number to an infinite collection? Before the nineteenth century mathematicians believed the answer to this question to be "no." That this was the case followed from two basic principles:

> (1) the axiom in Euclid's *Elements* that *the whole is always greater than the part.*

Here we shall take "greater than" in the strong sense of "having a greater number of members," rather than in the trivial sense of "having members not included in the part."

To state the second principle, we need to introduce the concept of the *equivalence* of collections. Two collections are said to be *equivalent* when their members can be put into *one-one correlation*, that is, correlated with each other in such a way that each member of the first collection is correlated with precisely one member of the second and vice-versa. Then the second principle is:

> (2) equivalent collections have the same cardinal number.

Principle (2) is familiar—in application at least—to everybody. In setting tables in a restaurant, if each place is set with just one knife and one fork, then the number of knives and the number of forks are the same, no matter how many there may be.

From (1) it follows that

> (3) if a given collection can be assigned a cardinal number, then its cardinal number must always exceed the cardinal numbers of any of its (proper) parts.

Now any *finite* collection can be assigned a cardinal number, and so satisfies (3). But in 1650 Galileo Galilei observed that if an infinite collection could be assigned a cardinal number, it would violate (3), and concluded that *infinite collections cannot be assigned cardinal numbers.* To be specific, he considered the infinite collection **N** of all the positive integers. If **N** could be assigned a cardinal number, then according to (3) its cardinal number would have to exceed those of its parts. Now consider the part **S** of **N** consisting of all square numbers 1, 2, 4, 9, ... The members of **N** can be one-one correlated with the elements of **S** by the pattern $1 \leftrightarrow 1$, $2 \leftrightarrow 4$, $3 \leftrightarrow 9$, ..., $n \leftrightarrow n^2$, ... So it follows from (2) that **N** and **S** have the *same* cardinal number, contradicting (3).

This argument is called *Galileo's paradox*. It embodies both the One/Many and Whole/Part oppositions.

Cantor realized that, if the idea of assigning cardinal numbers to infinite collections were to make sense, then *Euclid's axiom* (1) (with "greater than" understood in the strong sense above) *would have to be abandoned*. On the other hand, principle (2) could be retained as the basis for determining when the cardinal numbers of two collections are the same. In abandoning (1) and retaining (2) Cantor was making the bold, even revolutionary claim that there was *no difference in principle between finite and infinite sets*. This "collapsing" of the Finite/Infinite opposition played an essential role in Cantor's development of set theory.

PARADOXES OF THE INFINITE

The curious way in which Euclid's axiom is violated by infinite collections gives rise to a group of conundrums known collectively as *paradoxes of the infinite*. The most striking of these is the paradox of *Hilbert's Grand Hotel*, attributed to the great German mathematician David Hilbert. In this paradox, Hilbert manages a grand hotel, so grand, in fact, that it has an *infinite* number of rooms. Thus the hotel has a first, second, ..., n^{th}, ... room, *ad infinitum*. It is the height of the tourist season, and the hotel is fully occupied. A traveller seeking accommodation shows up at the hotel. "Alas," Hilbert tells her, "I have not a room to spare." But she is desperate, and at that point the gallant Hilbert has an idea. He asks the occupant of each room in the hotel to move to the next one; thus the occupant of room 1 is to move to room 2, that of room 2 to room 3, and so on up the line:

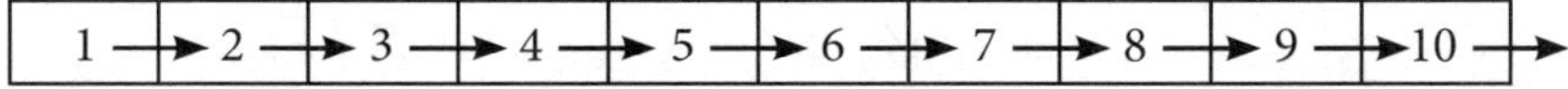

Once all the original occupants have shifted rooms in this way, they are accommodated—one to a room, as before—but this leaves the first room unoccupied and available to receive the new guest. In effect, an extra room has been "added," so that the set of rooms in Hilbert's Hotel is equivalent to the set of rooms along with one additional room.

Later, the fractious guest in room 1 complains that the bar of soap in his bathroom is too small. Unfortunately, the hotel has run out of soap bars. But Hilbert has recently engaged an enterprising new assistant manager,[1] who suggests the following expedient: ask each guest, apart from the guest occupying room 1, to hand his bar of soap to the guest occupying the room with number one less than his. This will leave the guest in room 1 with two bars of soap, and everybody else, as before with one bar of soap.

But then the guest in room 2 complains that it's unfair that his neighbour now has two bars of soap, while he has just one. In response the assistant manager asks each guest, apart from those in rooms 1 and 2, to hand his bar of soap to the guest occupying the room with number one less than his. This ensures that now the occupant of room 2 has two bars of soap.

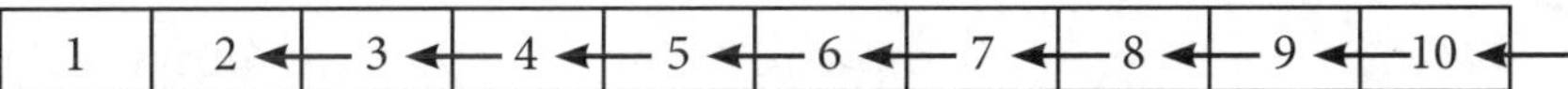

Clearly this procedure can be repeated indefinitely to supply each guest with any number of bars of soap.

The question immediately arises: *where did the "new" bars of soap come from?* But in fact there are no "new" bars of soap. Each "new" bar is just an "old" bar passed down from the next room. It is the infinite size of the collection of bars which produces the paradoxical illusion that new bars have somehow been "created."

Hilbert's hotel paradox has a sequel. One day an *infinite* congregation of tourists seeking accommodation descends on the hotel, which the hotel's popularity has again caused it to be full. Now Hilbert has foreseen just such an eventuality, and has accordingly redesigned his hotel in such a way as to make each room directly adjacent to the room with double the number, as follows:[2]

1	2	4	8	16	...
3	6	12	24	48	...
5	10	20	40	80	...
7	14	28	56	112	...
9	18	36	72	144	...
...	...	...	...	...	...

1 This role should naturally be played by José A. Bernardete, the originator of this version of the Hilbert hotel paradox.

2 Here the room in row *m*, column *n* is assigned the number $2^{n-1}(2^{m-1})$.

Hilbert now requests each guest to move to the room with double the number of the one he first occupied: thus the occupant of room 1 is to proceed to room 2, that of room 3 to room 6, etc. Since each room is now adjacent to the room with double the number, each guest can merely step into the next room, as in the case when just one new guest had to be accommodated. The result is again to leave all the original guests housed, only now each member of the infinite set of rooms bearing *odd numbers* has become vacant. Thus each new guest can be accommodated: the first in room 1, the second in room 3, the third in room 5, etc. So every guest, old and new, ends up happy.

This procedure, a version of Galileo's paradox, can be repeated indefinitely, enabling an infinite number of infinite assemblies of tourists to be put up.

Hilbert's hotel paradox shows that infinite sets have intriguingly counterintuitive properties, but not that they are *self-contradictory*. Indeed if we suppose that the physical universe contains infinitely many stars—as was actually the case in Newtonian cosmology—then the universe itself can assume the role of "Hilbert's Hotel," with the stars (or orbiting planets, assuming these exist) serving as "rooms." It seems not to have struck Newton's contemporaries that his infinite cosmos is thereby subject to Galileo's paradox. Newton's cosmology posed problems of a purely *physical*—and so presumably more pressing—nature, for example Olbers' (and others') problem of the dark sky at night and the problem, pointed out by H. Seeliger at the end of the nineteenth century, that masses in Newton's cosmos would be subject to arbitrarily large gravitational forces.

The *Tristram Shandy paradox*, due to Bertrand Russell, is another nice example of a paradox of the infinite. Tristram Shandy (the hero of a novel by the eighteenth-century English writer Laurence Sterne) writes his autobiography so slowly that it takes him one year to inscribe the events of a single day. If he is mortal he will clearly never finish his autobiography. But were he to live forever, then—despite the fact that he would still never actually complete his work—no part of his autobiography would remain unwritten, for to each day of his life there would be a corresponding year devoted to that day's description. This remains the case despite the fact that the interval between the day and the corresponding year grows indefinitely large with time.

UNCOUNTABLE INFINITIES

We have already introduced the idea of the *equivalence* of collections. This concept plays an essential role in Cantor's set theory. It gives precise expression to the idea of two sets *having the same size*. Equivalence for *finite* sets corresponds to the usual concept of *equality of cardinal number*, since two finite sets have the same number of elements precisely when the elements of the two sets can be put into one-to-one correspondence, and so the "size" of a finite set is completely determined by the number of its members. This is just an extension of the idea of counting, because in counting the elements of a finite set we place them in one-one correlation with a set of number symbols 1, 2, 3, ..., n, ... Cantor's idea was to extend the concept of equivalence to *infinite* sets, so as to enable them to be compared in size.

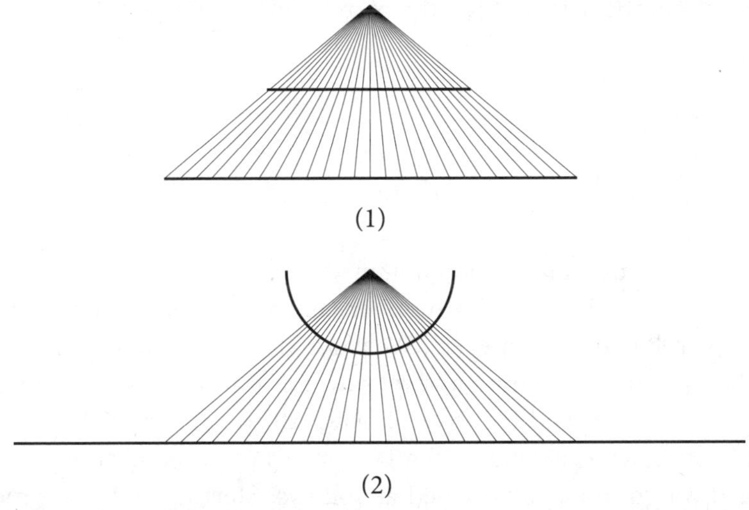

(1)

(2)

For infinite sets the idea of equivalence can lead to seemingly paradoxical results. For example, the sets of points on any pair of line segments are equivalent, even if the line segments are of different lengths. This can be seen by taking two arbitrary line segments and projecting from a point, as in figure (1) above. By bending one of the segments into a semicircle, as in figure (2), a similar procedure shows that this equivalence obtains even if one of the segments is replaced by the entire line.

One of Cantor's first discoveries in his charting of the infinite was that the set of *rational numbers*, the fractions—is actually *equivalent* to the set of natural numbers. The set of rational numbers contains the infinite set of natural numbers and so is itself infinite. It may strike one as paradoxical that the densely arranged set of rational numbers should be equivalent to its discretely arranged subset of natural numbers. After all, the (positive) rational numbers cannot be discretely arranged in *order of magnitude.* Starting with 0 as the first nonnegative rational, we cannot even choose a second "next larger" rational because, for every such choice, there will always be one smaller. Cantor observed that, if we *disregard* the relation of magnitude, it *is* possible to arrange all the positive rational numbers in a sequence $r_1, r_2, \ldots$ similar to that of the natural numbers. Such a sequential arrangement of a set of objects is called an *enumeration* of it, and the set is then called *enumerable.* By exhibiting such an enumeration, Cantor established the equivalence of the set of positive rational numbers with the set of natural numbers, since then the correlation

$$\begin{array}{cccc} 1 & 2 & 3 \ldots & n \ldots \\ \updownarrow & \updownarrow & \updownarrow & \updownarrow \\ r_1 & r_2 & r_3 \ldots & r_n \ldots \end{array}$$

is one-one.

How is this enumeration obtained?

Every positive rational number can be written in the form p/q, with p and q positive integers. We place all these numbers in an array, with p/q in the pth column and q^{th} row: for example, 5/7 is assigned to the fifth column and the seventh row. All the positive rational numbers may now be arranged in a sequence along a continuous "zigzag" line through the array defined as follows. Starting at 1, we proceed to the next entry on the right, obtaining 2 as the second member of the sequence, then diagonally down the left until the first column is reached at entry 1/2, then vertically down one place to 1/3, diagonally up until the first row is reached again at 3, across to 4, diagonally down to 1/4, and so on, as shown in the figure below. Proceeding along this zigzag line we generate a sequence 1, 2, 1/2, 1/3, 2/2, 3, 4, 3/2, 2/3, 1/4, 1/5, 2/4, 3/3, 4/2, 5, ... containing all the positive rationals in the order in which each is met along the line. In this sequence we now delete all those numbers p/q for which p and q have a common factor, so that each rational number will appear exactly once and in its

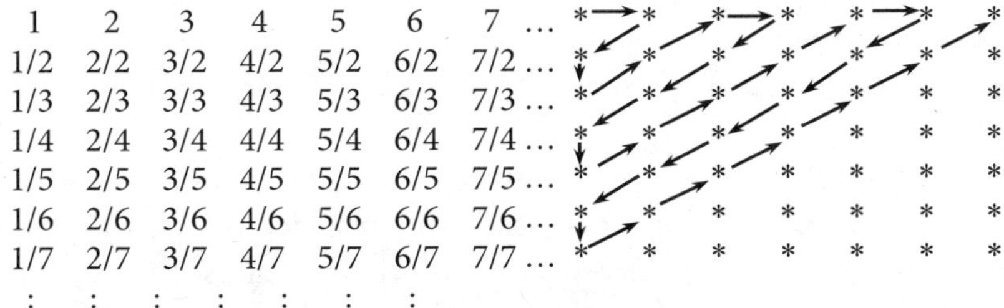

simplest form. In this way we obtain a sequence 1, 2, 1/2, 1/3, 3, 4, 3/2, 2/3, 1/4, 1/5, 5, ..., in which each rational occurs exactly once. This shows that the set of positive rational numbers is enumerable, and it is easy to deduce from this that the set of all rational numbers—positive, negative, or zero—is also enumerable. If r_1, r_2, ... is an enumeration of the positive rational numbers, then the sequence

$$0, r_1, -r_1, r_2, -r_2, \ldots$$

is an enumeration of all the rational numbers.

The fact that the rational numbers are enumerable might lead one to surmise that *any* infinite set is enumerable, thus bringing our tour of the infinite to a speedy conclusion. Nothing could be further from the truth. For Cantor also showed that the set of all *real* numbers, rational and irrational, is *not enumerable*, or *uncountable*. In other words, the collection of real numbers is a radically different, a *larger*, infinity than the collections of natural or rational numbers. The idea of Cantor's proof of this fact is to exhibit, for any given enumerated sequence s_1, s_2, ... of real numbers, a new real number which is *outside* it. As an immediate consequence, no given sequence s_1, s_2, ... can enumerate all real numbers. Therefore their totality is uncountable.

Any real number may be presented as an infinite decimal of the form

$$N.\, a_1 a_2 a_3 \ldots$$

where N denotes the integral part and the small letters the digits after the decimal point. For example ⅓=0.3333333... and $\pi = 3.1415926535\ldots$. Now suppose we are given an enumerated sequence s_1, s_2, s_3, ... of real numbers, with

$s_1 = N_1.\ a_1a_2a_3....$
$s_2 = N_2.\ b_1b_2b_3....$
$s_3 = N_3.\ c_1c_2c_3....$

We proceed to construct, using what has become known as *Cantor's diagonal process*, a new real number which we show does not occur in the given sequence. To do this we first choose a digit a that differs from a_1 and is neither 0 nor 9 (to avoid possible ambiguities which may arise from equalities like 0.999...= 1.000...), then a digit b different from b_2 and again unequal to 0 or 9, similarly c different from c_3 and so on down the "diagonal" of the above array. (For example, we might choose a to be 1 unless a_1 is 1, in which case we choose a to be 2, and similarly down the line for all the digits b, c, d, ...) Now consider the infinite decimal

$$x = 0.abcd\ ....$$

This new real number x then differs from any one of the numbers s_1, s_2, s_3, ... It cannot be equal to s_1 because the two differ at the first digit after the decimal point; it cannot be equal to s_2 because these two differ at the second digit after the decimal point, and so on. Accordingly x does not occur in the given sequence. So no given sequence can enumerate the totality of real numbers, which is therefore uncountable.

For set theory the importance of Cantor's theorem stems from the fact that it reveals the presence of at least *two* types of infinity: the enumerable infinity of the set of natural numbers and the uncountable infinity of the set of real numbers. We now extend the use of the term "cardinal number" to *arbitrary*—even *infinite*—sets by saying that two equivalent sets, whether they are finite or infinite, have the *same cardinal number.*[3] If A and B are finite this reduces to the statement that they have the same number of elements. Let us term *infinite* the cardinal number of an infinite set. Just as in the case of the set of rooms in Hilbert's Hotel, an infinite set is equivalent to the set obtained by

3 Strictly speaking, we have only introduced the phrase "have the same cardinal number" as a synonym for the term "equivalent." We have not specified what sort of thing the cardinal number of an infinite set actually *is*. In Cantor's theory a cardinal number is defined as a certain kind of ordered set, but the definition is too complicated to be given here.

adding one additional element, so that adding that one element does not change its cardinal number. This means that *an infinite cardinal number* **k** *satisfies the "paradoxical" equation*

$$\mathbf{k} + 1 = \mathbf{k}.$$

Infinite cardinal numbers are distinguished from ordinary integers precisely through the fact that they, and only they, satisfy this paradoxical equation.

In the same way, the fact that an infinite number of new guests can be accommodated in Hilbert's Hotel indicates that an infinite cardinal number **k** also satisfies the equation

$$\mathbf{k} + \mathbf{k} = \mathbf{k}.$$

Of course, zero is the only *finite* number which satisfies this equation.

Let us say that the cardinal number of a set *A does not exceed* that of a set *B* if *A* is equivalent to *B* itself, or a part of it. Let us also say that *A has* a *greater cardinal number* than a set *B* if *B* is equivalent to a part of *A*, but *A* is not equivalent to *B* or any of its parts. Since the set of natural numbers is a part of the set of real numbers, while, as we have seen, the latter is not equivalent to the former nor to any of its parts, it follows that *the set of real numbers has a greater cardinal number than the set of natural numbers.* Cantor went further and established the general fact that, *for any set A, it is possible to produce another set B with a greater cardinal number.* Let us define a *subset* of a set *A* to be a set which is a part of *A*, or *A* itself. The set *B* is chosen to be the *power set* of *A*, that is, the set of all subsets of *A*, which includes *A* itself and the empty subset ∅ containing no elements at all. We write **P**(*A*) for the power set of *A*. Thus each element of **P**(*A*) is *itself* a set, comprising certain elements of *A*. Now suppose that **P**(*A*) were equivalent to some subset of *A*, that is, there is a one-one correlation between the elements of *A* (or a subset of it) and the elements of **P**(*A*), i.e., with the subsets of *A*. We show that this is impossible by exhibiting a subset *T* of *A*, i.e., an element of **P**(*A*), which *cannot* have any element of *A* correlated with it. We obtain *T* as follows. For any element *x* of *A* which is correlated with an element of **P**(*A*) that is, with a subset of *A*, we have two possibilities: either *x* is a member of its correlated subset, or it is not. We define *T* to be the subset of *A* consisting of all correlated elements

x of A such that x is *not* an element of its correlated subset. We now show that T cannot not be correlated with any element of A. For suppose that T were correlated with some specific element a of A. Then, since T consists of all elements x of A for which x is not an element of its correlated subset, it follows, in particular, taking x to be a, that

> (1) a is an element of T exactly when a is not an element of its correlated subset.

But since T has been assumed to be the subset correlated with a, it follows from (1) that

> a is an element of T exactly when a is not an element of T.

This is a contradiction. We conclude that our original supposition was incorrect, and it follows that T cannot be correlated with any element of A.

Cantor's argument can also be formulated in a way similar to the bibliographic version of Russell's paradox described earlier in this chapter. Here the claim is: *for any set A*, no *correlation of elements of A with subsets of A—that is, with elements of* **P**(*A*)*—can be one -one*. In other words, any such correlation must be ambiguous in that there exists an element of A correlated with two *different* subsets of A. This is proved in the same way as for the bibliographies. Thus suppose we are given a correlation of elements of A with subsets of A. (The elements of A may then be thought of as analogous to *titles*; the subsets of A as analogous to *bibliographies;* the relation of membership between elements of A and subsets of A as analogous to the relation of a bibliography *listing* a title; and finally the given correlation of elements of A with subsets of A as analogous to the assignment of titles to bibliographies.) We define the subset U of A as follows. U is to consist of all those elements a of A for which a is correlated with a subset of *A not having u as a member*. Now let u be the element of A correlated with U. Then *u is an element of U*. For suppose u is not an element of U. Then u has the property that it is correlated with a subset of A, namely U, which does not have u as an element. That being the case, u satisfies the condition for being an element of U. Thus u has to be an element of U. It follows that u must be correlated with a subset V of A which does not have u has a member. Since u is a member of U, but not of V, U and V are different. Yet they are both correlated with u.

Both arguments establish the impossibility of setting up a one-one correlation between the elements of A, or one of its subsets, and those of its power set $\mathbf{P}(A)$. On the other hand, the correlation $x \leftrightarrow \{x\}$, where, for each element x of A, $\{x\}$ denotes the subset of A whose sole element is x, is a one-one correlation between A and the subset of $\mathbf{P}(A)$ consisting of all one-element subsets of A. It follows from the definition that $\mathbf{P}(A)$ has a greater cardinal number than A.

Thus Cantor showed that, just as there is no greatest integer, so there is no greatest infinite cardinal number. It follows that the infinite cardinal numbers form an ascending sequence which, like the sequence of integers, continues forever. Cantor used the symbol "ℵ"—*aleph*, the first letter of the Hebrew alphabet—to denote the members of this sequence, which is written $\aleph_0, \aleph_1, \aleph_2, \ldots$. Its first member, $\aleph_0$, is the cardinal number of the set of integers.[4]

As we have seen, the sets of points on any pair of line segments are equivalent, from which it follows that the cardinal number of the set of points on a line segment is independent of its length. It is a surprising, almost paradoxical, fact that, the cardinal number of the set of points in a geometric figure is independent of its *number of dimensions*. Thus, for example, we can show that the cardinal number of the set of points in a square is no greater than that of the set of points on one of its sides. This is done by setting up the following correlation.

If (x, y) is a point of the unit square (i.e., with $0 \leq x, y \leq 1$), x and y may be written in decimal form as

$$x = 0.a_1a_2a_3a_4\ldots.$$
$$y = 0.b_1b_2b_3b_4\ldots.$$

To the point (x, y) we then assign the point

$$z = 0.a_1b_1a_2b_2a_3b_3a_4b_4\ldots.$$

on the bottom side of the square, obtained by interlacing the two decimals corresponding to x and y. Clearly different points (x, y), (x', y') in the square are correlated with different points z, z' on the side, so that we have a one-one correlation between the set of points in the

4 '$\aleph_0$' is pronounced as 'aleph naught' or 'aleph zero' or 'aleph null'. '$\aleph_1$', '$\aleph_2$' ... are pronounced as 'aleph one', 'aleph 2', ...

square and a subset of the set of points on the side. Thus the cardinal number of the set of points in the square does not exceed the cardinal number of the set of points on a side. A similar argument shows that the cardinal number of the set of points in a cube is no greater than that of the set of points on an edge. By refining these arguments slightly it can be shown that the cardinal numbers of these various sets actually *coincide*.

These conclusions seem to contradict the intuitive idea of dimension. However, it should be noted that the correspondence between the points in a square and the points on a side we have set up is "unnatural" in not being *continuous*. This means that if we travel continuously along the points of the segment from 0 to 1, the corresponding points in the square do not form a continuous curve, but instead appear as a random, discontinuous pattern. It can in fact be shown that there are no continuous one-to-one correspondences between spaces of different dimensions. The dimension of a set of points in fact depends primarily on the way its elements are distributed in space, rather than on its cardinal number.

We have seen that the cardinal number of the set of real numbers—let us denote it by **c**—exceeds the cardinal number $\aleph_0$ of the set of integers. It is natural to ask: by *how much* does **c** exceed $\aleph_0$? That is, *how many* cardinal numbers are there (strictly) between $\aleph_0$ *and* **c**? The simplest possible answer is: *none*. The conjecture that this is the correct answer is called the continuum hypothesis. Since $\aleph_1$ *is by definition the next largest cardinal number above* $\aleph_0$, this hypothesis may be succinctly stated in the form

$$c = \aleph_1.$$

Cantor firmly believed in its correctness and expended much effort in attempting to prove it. His efforts were unsuccessful. It was not until 1963 that the American mathematician *Paul J. Cohen* showed that the continuum hypothesis *cannot* be proved from the principles on which set theory is built.[5] Later it was shown that **c** could coincide with virtually *any* cardinal number not less than $\aleph_1$. *The principles of set theory simply do not fix an exact value for c in terms of its position in the scale*

5 The Austrian logician Kurt Gödel showed in 1938 that the continuum hypothesis is consistent with these principles.

of alephs. Thus the status of the continuum hypothesis in set theory bears a certain resemblance to that of Euclid's fifth postulate in geometry—the "parallel postulate"—discussed in the next chapter. Just as the fifth postulate can be denied so as to produce a non-Euclidean geometry, so the continuum hypothesis can be denied so as to produce a "non-Cantorian" set theory.

SET-THEORETIC ANTINOMIES

Infinite sets are not contradictory in themselves. But as originally formulated set theory *did* contain contradictions (also called *antinomies*). These result not from admitting infinite collections *per se*, but rather by allowing the formation of pluralities, such as Russell's collection, which cannot, on pain of contradiction, be treated as "ones." It was in fact not the Finite/Infinite opposition, but rather the One/Many opposition, which led set theory to inconsistency in its early phases. Three antinomies made their appearance. One of these is Russell's paradox. The other two, which appeared before Russell's paradox, are *Cantor's antinomy of the greatest cardinal number*, and *Burali-Forti's antinomy of the greatest ordinal number*.

Cantor's antinomy arises from the fact that there is no greatest cardinal number. Let us recall Cantor's definition of set:

> By a set we understand every collection to a whole of definite, well-differentiated objects of our intuition or thought.

Cantor thought that, since sets are made up of distinct individuals, it always makes sense to ask whether two sets can be put into one-one correlation, so that any set can be assigned a cardinal number. Now consider the collection V of *all* sets. Then if V were a set it would have a cardinal number—call it Ω. But in that case Ω must be the greatest cardinal number. For let $\mathbf{k}$ be any cardinal number, and let A be a set with cardinal number $\mathbf{k}$. Then A, as a set, is part of V, and so its cardinal number cannot exceed that of V. That is, $\mathbf{k}$ cannot exceed Ω. Since $\mathbf{k}$ was arbitrary, it follows that no cardinal number can exceed Ω, so that Ω would have to be the greatest cardinal number, contrary to the fact that—as we have seen—there is no greatest cardinal number. The only way out of this contradiction is to infer that $\mathbf{V}$ is not a set, that is, it has no cardinal number. Accordingly **V** provides another instance of a Many not treatable as a One.

Another way of showing that **V** is not a set is to invoke the idea behind Russell's paradox. To do this we need to assume the following principle:

> **Principle of Parts**. Any part of a Many treatable as One is itself treatable as One.

Put in terms of sets, the Principle of Parts asserts that *any subcollection of a set is itself a set.*

Now it follows from this principle that, for any set of sets **S**, there is a set **S'** which is not an element of **S**. For define **S'** to be the collection of all members of **S** which are not elements of themselves. We infer from the Principle of Parts that **S'**, as a part of the set **S**, is itself a set. Suppose now that **S'** were an element of **S**. Then from its very definition it follows immediately that **S'** is an element of itself just when it is not an element of itself. We conclude that **S'** cannot be an element of **S**.

We infer from this that the collection **V** of all sets cannot be a set, since by definition there is no set which fails to be a member of **V**. Put differently, each set **S** is extensible in that it "generates" another set **S'** which is not a member of it; since **V** is not so extensible, it cannot be a set.

A similar antinomy arises in connection with finite sets. Recall that we defined a set to be finite if its elements can be counted by natural numbers to termination, yielding the (natural) number of elements of the set. Next, note that there is no largest natural number. Now consider the class **F** of all finite sets. Then **F** cannot be finite, for if it were, it would have a definite (natural) number of elements, and clearly this would have to be the greatest natural number. This is a contradiction, and so **F** cannot be finite.

The second of these set-theoretic antinomies, that of Burali-Forti, concerns infinite ordinal numbers. Cantor had extended the well-ordered[6] sequence of finite ordinal numbers first, second, third, etc. to a similarly well-ordered sequence of infinite ordinal numbers, and he had shown that corresponding to each well-ordered set there is a unique ordinal number. He had also shown that, just as there is no largest finite ordinal number, there is no greatest infinite ordinal. Now consider the collection **W** of all (finite and infinite) ordinals. If **W** were

6 Formally, an ordered set is *well-ordered* if any non-empty subset has a least element.

a set, then because it is well ordered, it would have an ordinal assigned to it which would have to be the greatest ordinal. This contradiction shows that, like **R** and **V**, the collection **W** cannot be a set—like them, it is a Many not treatable as a One.

THE AXIOM OF CHOICE

A principle of set theory which has generated much dispute is the celebrated *axiom of choice*. Introduced in 1904 by the German mathematician Ernst Zermelo, in its simplest form it asserts that, if we are given any collection **S** of sets, each of which has at least one element, and no two of which have an element in common, there is a collection M consisting of exactly one element from each set in **S**. Thinking of each set in **S** as a "grab bag," what the axiom of choice asserts is that from each "bag" a member can be "grabbed" and then "deposited" into the initially empty "bag" M. There is no problem in assembling M when there are only finitely many sets in **S**. Even if **S** comprises infinitely many sets, M can, in principle, be assembled provided we possess a *definite rule* for choosing a member from each set in **S**. The difficulty arises when **S** contains infinitely many sets, but we have *no rule* for selecting a member from each. In this situation, how can the procedure be justified of making infinitely many arbitrary choices, and forming a collection from the result?

The difficulty here is well illustrated by the following scenario, due to Bertrand Russell.[7] Wooster, a billionaire, possesses an infinite number of pairs of shoes, and an infinite number of pairs of socks. One day, he summons his valet Jeeves and requests him to select one shoe from each pair. Accustomed to receiving precise instructions, Jeeves asks for details as to how to go about doing this. Wooster, in a fit of inspiration, suggests that the left shoe be chosen from each pair. Jeeves duly performs this operation, and presents his employer with the resulting pile of shoes. The following day Wooster rings for Jeeves to select one *sock* from each pair. When Jeeves asks Wooster how *this* operation is to be carried out, the latter is at a loss for a reply. Now Jeeves, sharper than his employer, has grasped that, unlike pairs of shoes, there is no intrinsic way of distinguishing one sock of a pair from the other. In other words, the selection of the socks must be truly arbitrary. So, in order to meet his employer's requirements, Jeeves simply select a sock *at random*.

---

7 Here presented using characters created by P.G. Wodehouse.

The collection of socks assembled at random by Jeeves is just that—a *random collection*, with no rule or common property uniting its members. The axiom of choice asserts no more than the existence of collections whose members may be thought of as "arising" by random selection. Put this way, the axiom of choice seems reasonable. Nevertheless, it does have some strange consequences.

One of the oddest of these is the so-called *paradoxical decomposition of the sphere*,[8] formulated in 1924 by the Polish mathematicians Stefan Banach and Alfred Tarski. The "paradox" here is that, assuming the axiom of choice, a solid sphere can be decomposed, or "cut up" into finitely many pieces which can themselves be reassembled exactly to form *two* solid spheres, *each* of the same size as the original! More precisely, any solid sphere can be can be decomposed into a finite number of non-overlapping subsets,[9] which can be reassembled by translation and rotation (i.e., without changing their shape) in such a way as to produce two solid spheres, each of the same size as the original. It should be noted that the "pieces" are not solids in the usual sense, but infinite scattered collections of points.

A second version of the paradox is that, given any pair of solid spheres, either one of them can be cut up into finitely many pieces which can be reassembled to form a solid sphere of the same size as the other. Thus, a sphere the size of the sun can be cut up and reassembled to form a sphere the size of a pea, and vice-versa!

The counterintuitive nature of all this is illustrated by the fact that, if it were true of material objects, the principle of conservation of mass would be violated in a spectacular way. The first version of the paradox could lead to the following scenario. Someone buys from a novelty store (Messrs. S. Banach and A. Tarski, *props.*) one of those intriguing 3-dimensional oriental jigsaw puzzles—wooden spheres which after disassembly are to be reassembled in their original spherical form. The purchaser is astounded to find that upon reassembling the pieces making up the original sphere, he now has *two* solid spheres of the same size as the original. The second version of the paradox could lead a sufficiently dextrous (and unscrupulous) person to purchase a

8 See S. Wagon, *The Banach-Tarski Paradox* (Cambridge: Cambridge University Press, 1985).

9 Even more surprisingly, it has been shown that no more than five such subsets are required.

gold ingot, melt it down, recast it in spherical form[10] (one thinks of the gold Eiffel towers in the film *The Lavender Hill Mob*) and, applying the Banach-Tarski technique, section it into a number of pieces which he finally reassembles into a gold sphere of a size requiring Fort Knox to accommodate it.

Of course, in stating these paradoxes, the phrase "cut up" is to be taken in a metaphorical, not practical, sense; but this does not detract from their strangeness. Bizarre as they are, however, unlike Russell's paradox they do not constitute outright contradictions. Actually, sphere decompositions of this kind are possible in set theory only because continuous geometric objects have been analyzed into discrete sets of points to which the axiom of choice can then be applied so as to rearrange them in an arbitrary manner. In the end this boils down to another instance of the opposition between the Continuous and the Discrete.

The perplexities surrounding the emergence of set theory are collectively designated by historians of mathematics as the *third crisis* in the foundations of mathematics (the two previous being the Pythagorean discovery of incommensurables, and the shaky state of the foundations of the calculus in the seventeenth and eighteenth centuries). Attempts to resolve this crisis took several different forms, all of which involved subtle analysis of the nature of mathematical concepts and reasoning.

The purely technical difficulties in set theory were overcome when, in the first few decades of the twentieth century, set theory was *axiomatized* (by Zermelo and others) in such a way as to circumvent contradictions such as Russell's paradox by suitably restricting the ways in which sets could be formed. Any residual doubts concerning the acceptability of the axiom of choice were dispelled in 1938 when Gödel established its consistency with respect to the remaining axioms of set theory. These developments enabled the majority of mathematicians to accept set theory as providing an adequate foundation for their work. Mathematicians find set theory acceptable not solely for the practical reason that it enables mathematics to be *done*, but

10 Strictly speaking, this part of the procedure is unnecessary because the Banach-Tarski paradox applies, *mutatis mutandis*, to any region of 3-dimensional space which is topologically equivalent to (i.e., continuously deformable into) a solid sphere.

also because it is consonant with the unspoken belief of the majority that, like sets, mathematical objects actually *exist* in some sense and mathematical theorems express truths about these objects. This is a version of the philosophical doctrine of *Realism*, also termed, with less accuracy, *Platonism*.

Hermann Weyl, the great twentieth-century German mathematician, has written:

> 'Mathematizing' may well be a human creative activity, like music, the products of which not only in form but also in substance are conditioned by the decisions of history and therefore defy complete objective rationalization.

This is a perceptive observation; nevertheless, the attempt to explicate the mathematical infinite, and so to grasp the ultimate nature of mathematics, has borne much fruit and will no doubt long continue to be a source of inspiration.

Chapter III

The Strange Universe of Non-Euclidean Geometry

HYPERBOLIC GEOMETRY

By far the most influential mathematical text in history is the celebrated *Elements* of *Euclid*, written about 300 BCE. This definitive treatise on geometry appears to have quickly and completely superseded all previous works of its kind: in fact, no trace whatsoever remains of its predecessors. With the exception of the Bible, no work has been circulated more widely or studied more assiduously, and no work, without exception, has exercised a greater influence on scientific thinking. Over a thousand editions of the *Elements* have appeared since the first was printed in 1482, and for more than two millennia it dominated the teaching of geometry. For centuries the idea of a geometric system not absolutely in accordance with Euclid's would have been regarded as nonsensical.

As it has come down to us, the *Elements* opens with 23 definitions, of which the following may be taken as typical:

1. A *point* is that which has no part.
2. A *line* is breadthless length.
3. A *straight line* is a line which lies evenly with the points in itself.
8. A *plane angle* is the inclination to one another of two lines in a plane which meet one another and do not lie in a straight line.
10. When a straight line set up on a straight line makes adjacent angles equal to one another, each of the equal angles is *right*, and the straight line standing on the other is called *perpendicular* to that on which it stands.
15. A *circle* is a plane figure enclosed by one line such that all the straight lines falling upon it from one point among those lying within the figure are equal to one another.
23. *Parallel* straight lines are those which, being in the same plane and being produced indefinitely in both directions, do not meet one another in either direction.

The work continues with five *axioms* and five geometric *postulates*:

A1. Things which are equal to the same thing are equal to each other.
A2. If equals be added to equals, the wholes are equal.
A3. If equals be subtracted from equals, the remainders are equal.
A4. Things which coincide with one another are equal to one another.
A5. The whole is greater than the part.

P1. It is possible to draw a straight line from any point to any other point.
P2. It is possible to produce a straight line indefinitely in that straight line.
P3. It is possible to describe a circle with any point as centre and with a radius equal to any finite straight line drawn from the centre.
P4. All right angles are equal to one another.
P5. If a straight line intersects two straight lines so as to make the interior angles on one side of it together less than two right angles, these straight lines will intersect, if indefinitely produced, on the side on which the angles are together less than two right angles.

These postulates were put forward as a body of assertions about lines and points conceived as lying in physical space whose correctness would be immediately obvious to everyone. The correctness of the fifth, or *parallel* postulate, however, is by no means obvious. This postulate is now usually stated in the equivalent form (popularized by the eighteenth-century Scottish mathematician John Playfair) that *through any point not on a given line one and only one line may be drawn parallel to the given line*. The striking feature of this postulate is that it makes an assertion about the *whole extent* of a straight line, imagined as being extended indefinitely in either direction; for two lines are *defined* to be parallel if they never intersect, however far they are produced. Now of course there are many lines through a point which do not intersect a given line within any prescribed finite distance, however large. Since the maximum possible length of an *actual* ruler, thread, or even light ray visible through a telescope is certainly finite, and since within any finite circle there are infinitely many straight lines passing through a given point and not intersecting a given straight line, it follows that the postulate can never be verified—or even refuted—by experiment. On the other hand, all the other postulates of Euclidean geometry—with the exception of the second postulate asserting that straight lines can be indefinitely extended—have a *finite* character in that they deal with bounded portions of lines and planes. The fact that the parallel postulate is not experimentally verifiable, while the remaining postulates are, suggested the idea of trying to *derive* it from the latter. For centuries, mathematicians strove without success to find such a derivation.

One of the first attempts in this direction was made by *Proclus* (fourth century BCE), who tried to dispense with the need for a special parallel postulate by the ingenious expedient of *defining* the parallel to a given line to be the locus of all points at a fixed distance from the line. Unfortunately, it then became necessary to show that the locus of such points is indeed a straight line! Since this assertion is actually *equivalent* to the parallel postulate, Proclus made no real advance here.

Not until 1733 was the first truly scientific investigation of the parallel postulate published. In that year there appeared the book *Euclides ab omni naevo vindicatis*—"Euclid Freed of Every Flaw"—by the Italian Jesuit Girolamo Saccheri (1667–1733). Without using the parallel postulate, Saccheri easily showed that if, in a quadrilateral

ABCD, the angles at *A* and *B* are right angles and sides *AD* and *BC* are equal, then the angles at *D* and *C* are also equal. There are then three possibilities: the angles at *C* and *D* are equal *acute, right, or obtuse* angles. Saccheri showed, using the remaining postulates of Euclidean geometry, that the case of obtuse angles is impossible. His aim was

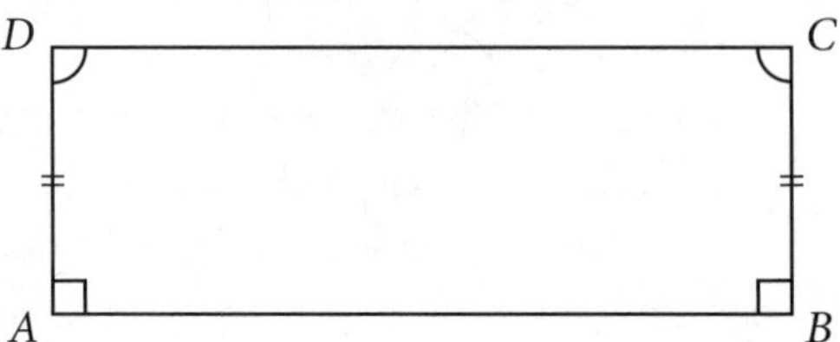

to demonstrate that the acute angle hypothesis also leads to a contradiction, thus leaving the right angle case, which is easily shown to be equivalent to the parallel postulate. Saccheri's method was thus to assume the acute angle hypothesis, together with the postulates of Euclidean geometry apart from the parallel postulate and attempt to derive a contradiction. A successful outcome would show the parallel postulate to be a consequence of the remaining ones. Remarkably, in the course of his investigations, Saccheri derived many of the theorems of what was later to become known as non-Euclidean geometry. Unfortunately, however, he completed his discussion by deriving an unconvincing "contradiction" involving a nebulous idea of "infinite element." Had he not been so eager to exhibit a contradiction here, but had rather admitted his inability to find one, Saccheri would today indisputably be credited with the discovery of non-Euclidean geometry.

It was the continual failure by mathematicians to find a proof of the parallel postulate that finally convinced them that it must be truly *independent* of the other postulates. Thus a self-consistent *non-Euclidean* geometry, in which the parallel postulate is *false*, became conceivable.

The Hungarian mathematician Janos Bolyai and the Russian mathematician Nikolai Ivanovich Lobachevsky were the first to publish, independently, in 1832 and 1829, respectively, detailed accounts of a system of non-Euclidean geometry.[1] This system, known as *hyperbolic*

1 Both were in fact anticipated by Gauss, who, however, fearing critical reaction—to which he referred as "the cries of the Boeotians"—never published his discoveries

(or Bolyai-Lobachevsky) geometry possesses certain curious—almost paradoxical—features that set it apart from Euclidean geometry in a most dramatic way, and which fully justify Bolyai's description of it as a "strange new universe."

To begin with, there is always *more than one straight line parallel to a given one passing through a given point outside it*. Let us examine this situation a little more closely. Calling the given line λ and the point *P*, drop perpendicular *PQ* to λ and let *m* be the perpendicular through *P* to *PQ*. Consider one ray *PS* of *m* and various rays between *PS* and *PQ*. Some of these rays, such as *PR*, will intersect λ, while others, such as *PY*, will not. As *R* recedes indefinitely on λ from *Q*, *PR* will approach a certain *limiting* ray *PX* that does *not* meet λ. The ray *PX* is "limiting" in the sense that any ray between *PX* and *PQ* intersects λ, whereas any other ray *PY* such that *PX* is between *PY* and *PQ*, will fail to do so. The ray *PX* is called the *left limiting parallel ray to λ through P*.

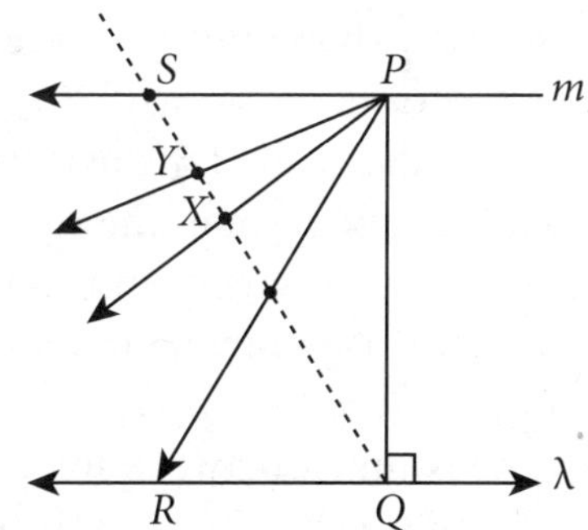

Similarly, there is a *right* limiting parallel ray *PX*' on the opposite side of *PQ*. These limiting rays are symmetrically situated about *PQ*. The angle *QPX* = *QPX*' is called the *angle of parallelism* at *P* with

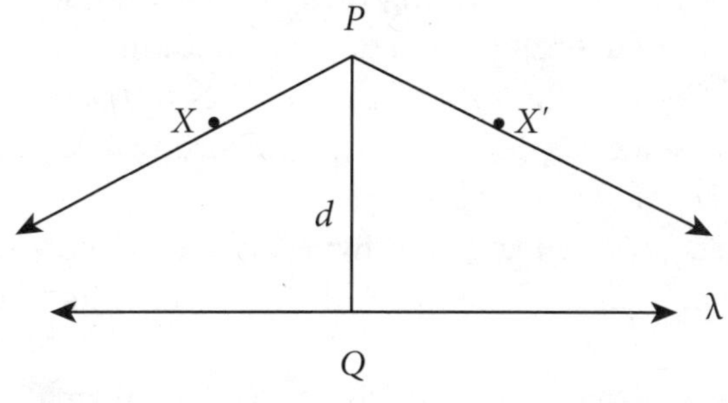

in non-Euclidean geometry.

respect to λ: the size of this angle depends only on the distance d from P to Q, and not on the identity of the particular line λ, nor on that of the particular point P. As the distance d varies, the angle of parallelism α takes on all values between 0 and 90 degrees: as P approaches Q, α approaches 90°, and as P recedes to infinity from Q, α approaches 0°.

Triangles also behave oddly in hyperbolic geometry. All triangles have angle sum *less than* 180°, so that, if we define the *defect* of a triangle to be the difference between 180° and the sum of its angles, the defect of any triangle is always positive. As a triangle shrinks, however, its defect becomes arbitrarily small. Another striking fact about triangles in hyperbolic geometry is that *all similar triangles are congruent.* (Two triangles are *similar* if their corresponding angles are the same.) This means that a given triangle *cannot be enlarged or shrunk without changing its shape.* A startling consequence is that a segment can be determined with the aid of an angle: for example, an angle (50°, say) of an equilateral triangle determines the length of the side uniquely. To put it more dramatically, hyperbolic geometry has an *absolute measure of length.* This contrasts starkly with Euclidean geometry. For while it shares with hyperbolic geometry the feature of possessing an absolute measure of angle in the form of the right angle, there can be no absolute measure of length in Euclidean geometry since there the geometric properties of figures are invariant under change of scale. In Euclidean geometry length is *relative.*

So in this respect the contrast between hyperbolic geometry and Euclidean geometry can be seen as an instance of the opposition between the Absolute and the Relative.

In hyperbolic geometry *all convex quadrilaterals have angle sum less than 360°*: in particular, there are no quadrilaterals containing four right angles; that is, *there are no rectangles or squares.* Since the customary system of measuring *area* is based on square units, this makes the task of defining area a ticklish affair. In fact the only reasonable way of defining the area of a triangle is to make it proportional to the defect. Since the defect can never exceed 180°, *there is an upper bound to the area of a triangle.*

A further curious feature of hyperbolic geometry is that the circumference of a circle of radius r always *exceeds* its Euclidean value $2\pi r$.

Although Lobachevsky actually showed, by formal methods, that hyperbolic geometry was consistent, this fact seems to have gone unrecognized at the time. The consistency of hyperbolic geometry was not in fact publicly affirmed until Arthur Cayley (1821–1895), Eugenio Beltrami (1835–1900), Felix Klein (1849–1925), and others constructed *models* for it, that is, interpretations under which all its postulates could be seen to be true. In Klein's model we take a fixed circle *C* in the Euclidean plane and interpret *point* as *point in C*, *line* as *line in C*, and, glancing at the figure below, *distance between P*

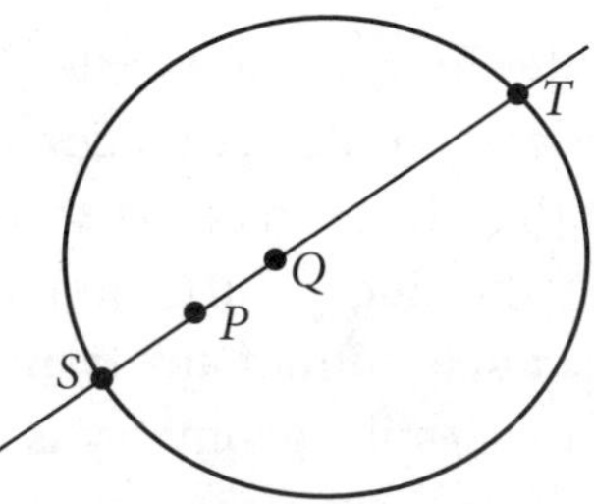

and Q as a certain function of *P* which becomes arbitrarily large as *P* approaches *S* or *Q* approaches *T*. This latter fact ensures that the "lines" in the model can be indefinitely "extended." In this model the parallel postulate obviously fails—there being many "lines" passing through a given point "parallel" to a given "line" in the sense that they fail to intersect *within C*. On the other hand the distance function can be chosen in such a way as to ensure that the other postulates of Euclidean geometry remain true.

Euclidean geometry may be thought of as the geometry of a flat surface—its *intrinsic geometry*. Hyperbolic geometry may be visualized as the intrinsic geometry of a curved, saddle-shaped surface.[2] Such a surface looks like this:

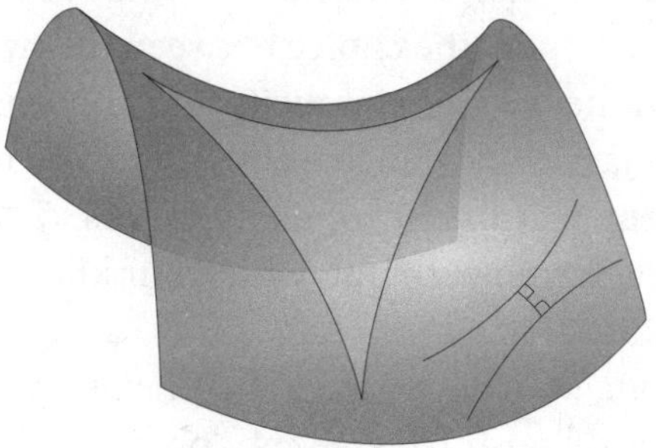

2 This shows that hyperbolic geometry also embodies the opposition between the Straight and the Curved.

On this surface a "straight line" between two points is a curve of least length—a *geodesic*—joining the points. In general these will be curved lines, not straight lines in the Euclidean sense, and there will be—as illustrated in the figure—divergent parallel lines. Also shown in the figure is a typical triangle on the surface. As the vertices of such a triangle are pulled farther and farther apart, its curved sides squeeze closer and closer together so that its area does not exceed a certain maximum.

An intelligent ant dwelling on a saddle-shaped surface would be able (in principle at least) to detect its curvature by making measurements which show that the surface geometry conforms with the hyperbolic (rather than the Euclidean) framework. For example, it would be found that the angle sum of any triangle is always less than 180°. The validation of hyperbolic geometry is an intrinsic feature of the surface which reflects its curvature.

Similarly, a three (or higher) dimensional space satisfying hyperbolic geometry can be conceived as possessing "curvature." In that case, it is meaningful to ask whether space in the *real world* is curved in the sense that its geometry is hyperbolic. In the nineteenth century the great German mathematician Carl Friedrich Gauss put this issue to the test by actually performing an experiment, using sextants, to determine whether the angle sum of a triangle is equal to, or less than, 180°. But although the result was, within the limits of experimental error, 180°, nothing was settled since, for small triangles (i.e., of terrestrial dimensions), the deviation from 180° might be so small as to be experimentally undetectable. That is, hyperbolic and Euclidean geometry, although differing *in the large*, may coincide so closely *in the small* as to be empirically equivalent. So far as *local* properties of space are concerned, then, the choice between the two geometries can be made solely on the basis of simplicity and convenience.

The revolutionary importance of the discovery of non-Euclidean geometry lay in the fact that it toppled Euclid's system as the immutable mathematical framework into which our knowledge of objective reality must be fitted.

The *mathematician* may regard a "geometry" as being defined by any consistent set of postulates about "points," "lines," etc. But the *physicist* will only find the result useful when the postulates in question accord with the behaviour of entities in the real world. Consider, for example, the statement *light travels in a straight line*. If this is regarded as the *physical definition* of a straight line, then the postulates of geometry must be chosen so as to correspond with the actual behaviour of light rays. In the late nineteenth century the great French mathematician Henri Poincaré imagined a world confined within the interior of a circle *C*, in which the velocity of light at each point is inversely proportional to the distance of the point from the circumference of *C* (for example, *C* could be made of glass of suitably varying refractive index). It can then be shown that light rays will assume the form of *circular arcs* perpendicular at their extremities to the circumference of *C*, and thus that hyperbolic geometry will prevail. Nevertheless, we can also arrange for *Euclidean* geometry to apply in this world: instead of regarding light rays as "non-Euclidean" straight lines, we simply take them to be Euclidean circles normal to *C*. Different geometries can describe the same physical situation, provided that physical entities (in the case just considered, light rays) are correlated with different notions in the geometries concerned.

RIEMANNIAN GEOMETRY

In both hyperbolic and Euclidean geometry it is tacitly assumed that *lines can be indefinitely extended*. But after Bolyai and Lobachevsky had revealed the possibility of constructing new geometries, it became natural to ask whether non-Euclidean geometries could be constructed in which "straight lines" are not infinite, but *finite* and *closed*. Such geometries were first considered in 1851 by the great German mathematician *Georg Friedrich Bernhard Riemann*. It turns out that geometries with closed finite lines can be constructed in a completely consistent way. An example is the intrinsic geometry of a surface of a sphere. In this geometry straight lines correspond to great circles on the surface. Since every pair of great circles intersect (in two points), in this geometry there are *no* "parallel lines" at all.

Riemann extended the idea of his geometry by considering a "space" consisting of an arbitrary curved surface, and defining a "straight line" between two points on the surface to be the curve of shortest length or *geodesic* on the surface joining the points. In this case the deviation of the geodesics from Euclidean straightness provides a measure of the curvature of the surface.

In his famous lecture of 1854, published as a paper in 1868, "On the Hypotheses which Lie at the Foundations of Geometry," Riemann introduced the idea of an intrinsic geometry for an arbitrary "space" which he termed a *multiply extended manifold*. Riemann conceived of a manifold as being the domain over which varies what he termed a *multiply extended magnitude*. Such a magnitude M he called *n-fold extended*, and the associated manifold *n-dimensional*, if n quantities—called *coordinates*—need to be specified in order to fix the value of M. For example, the position of a rigid body is a 6-fold extended magnitude because three quantities are required to specify its location and another three to specify its orientation in space. Similarly, the fact that pure musical tones are determined by giving intensity and pitch show these to be 2-fold extended magnitudes. In both of these cases the associated manifold is *continuous* in so far as each magnitude is capable of varying continuously with no "gaps." By contrast, Riemann termed *discrete* a manifold whose associated magnitude jumps discontinuously from one value to another, such as, for example, the number of leaves on the branches of a tree. Of discrete manifolds Riemann remarks:

> Concepts whose modes of determination form a discrete manifold are so numerous, that for things arbitrarily given there can always be found a concept ... under which they are comprehended, and mathematicians have been able therefore in the doctrine of discrete quantities to set out without scruple from the postulate that given things are to be considered as being all of one kind. On the other hand there are in common life only such infrequent occasions to form concepts whose modes of determination form a continuous manifold, that the positions of objects of sense, and the colours, are probably the only simple notions whose modes of determination form a continuous manifold. More frequent occasion for the birth and development of these notions is first found in higher mathematics.

Riemann observed that the sizes of parts of discrete manifolds can be compared by straightforward counting, and the matter ends there. In the case of continuous manifolds, on the other hand, such comparisons must be made by *measurement*. Measurement, however requires the positing a *unit of measurement*, that is of some magnitude—not a pure number—independent of its place in the manifold.

Riemann thought of a continuous manifold as a generalization of the three-dimensional space of experience, and refers to the coordinates of the associated continuous magnitudes as *points*. He was convinced that our acquaintance with physical space arises only *locally*, that is, through the experience of phenomena arising in our immediate neighbourhood. In particular, the *distance* between two points in a manifold is defined in the first instance only between points which are at *infinitesimal* distance from one another. This distance, known as the *metric*, is calculated according to a natural generalization of the distance formula in Euclidean space. In making this generalization Riemann allows for the possibility that the metric of the manifold or "space" may vary from point to point, just as the curvature of a surface may so vary.

Riemann also introduced the concept of *total curvature* of a manifold. In doing so his goal was to characterize Euclidean space, and, more generally, spaces which are *homogeneous* in that within them figures can be moved about without change of size or shape. Riemann's notion of curvature of a manifold is an intrinsic property of the manifold (or, more precisely, of its metric); it is not required to think of the manifold as being situated in some manifold of higher dimension. Riemann introduced manifolds of *constant curvature*: by definition, in these spaces all measures of curvature are equal and remain unchanged from point to point, so yielding the required homogeneity property.

If the curvature is positive we obtain a spherical space; it is zero, a Euclidean (flat) space, and if it is negative, a space resembling the inside surface of a torus.

In the final section of his paper, Riemann applied his ideas to the problem of determining the structure of *physical space*. He pointed out that, as regards physical space, *infinitude* must be carefully distinguished from *boundlessness*. For example, the surface of a sphere is finite but unbounded and, for all we know, the same may be true of physical space. In any case it is the boundlessness, rather than the infinitude of space that is required for manoeuvring in the external world, so that the former has a far greater empirical certainty than the latter. Moreover, if space has a constant positive curvature, however small, then it would take the form of a spherical surface and would accordingly be finite and boundless.

Riemann concluded his discussion with the following words, the last sentence of which proved to be prophetic:

While in a discrete manifold the principle of metric relations is implicit in the notion of this manifold, it must come from somewhere else in the case of a continuous manifold. Either then the actual things forming the groundwork of a space must constitute a discrete manifold, or else the basis of metric relations must be sought for outside that actuality, in colligating[3] forces that operate on it. A decision on these questions can only be found by starting from the structure of phenomena that has been confirmed in experience hitherto ... and by modifying the structure gradually under the compulsion of facts which it cannot explain.... This path leads out into the domain of another science, into the realm of physics.

Riemann was saying, in other words, that if physical space is a continuous manifold, then its geometry cannot be derived *a priori*—as claimed, famously, by Kant—but can only be determined by experience. In particular, and again in opposition to Kant, who held that the axioms of Euclidean geometry were necessarily and exactly true of our conception of space, these axioms may have no more than approximate truth.

Riemann's final sentence proved to be prophetic because in 1916 his geometry—*Riemannian geometry*—was to provide the basis for a landmark development in physics, Einstein's celebrated *General Theory of Relativity*, to be discussed in Chapter V. In Einstein's theory, the geometry of space is determined by the gravitational influence of the matter contained in it, thus perfectly realizing Riemann's contention that this geometry must come from "somewhere else," i.e., from physics.

3 Linking together.

Chapter IV

Puzzles and Paradoxes of Time Travel

Space and time are similar in that they are both taken to be continuous. There are no "gaps" in space. And while there may be gaps in one's memory—in *subjective* time—we are inclined to believe that there are no gaps in actual—*objective*—time. Objective time is, so to speak, the gapless "interval" between events. Nevertheless, space and time differ in certain fundamental respects. One is the fact that while locations in space form a three-dimensional whole, "moments" or "events" in time are ordered in a linear—a one-dimensional—pattern. Another is the fact that it is possible to move about—to travel—freely in space. This is so because space is *isotropic*—there are no privileged locations or directions in space. Time, or subjective time, at least, differs markedly from space in that it does possess privileged locations—known as *past, present, and future*—as well as a privileged direction—from *past* to *future*. Because of this, the very idea of "travel" through time bristles with difficulties.

The analysis of time travel offered here will rest chiefly on a fundamental law of logic: the *law of noncontradiction*. This asserts that, for any statement *p*, it cannot be the case that both *p* and not *p* are true. It is, in other words, logically impossible for a statement to be both true and false. In the presence of the law of noncontradiction, we shall see that the possibility of time travel has some truly starling consequences. These consequences will provide striking illustrations of the oppositions between the One and the Many, the Subjective and the Objective, and Chance and Necessity.

TIME TRAVEL INTO THE PAST: BRANCHING TIMELINES

Here is a possible time travel scenario. Suppose that in the year 2004 Tom Swift,[1] a 30 year old inventor, has, after a decade of effort, succeeded in constructing a machine capable of sending him into the past. He is convinced that his device will enable him to travel into the past and meet his younger self. Eager to put this to the test, he decides to travel 10 years into the past and present himself to his 20-year-old self at the very moment the latter first conceives the idea of designing a time machine. The 30 year old Tom recalls that this occurred when his younger self was sitting at his desk in his room at precisely 4 p.m. on Dec. 6, 1994. Accordingly at 4 p.m. on Dec. 6, 2004, the 30-year-old Tom sets the time machine's dial so as to arrive in his younger self's room at exactly 4 p.m. on Dec. 6, 1994. He presses the machine's button, and is returned to that place and moment. Emerging from the machine, the older Tom finds, as he had anticipated, the younger Tom sitting at his desk. The older Tom presents himself to the younger

1 Readers not acquainted with the name "Tom Swift" may be interested to learn that he was the main character of a popular series of American juvenile science fiction adventure novels (a number of which the present author read in his youth). First published in 1910, and continuing to appear for nearly a century, the series comprised upward of 100 volumes. The novels were translated into many languages, and sold more than 30 million copies worldwide. In most of the books, Tom Swift is presented as a youthful inventor of genius, creating the extraordinary inventions of the books' titles. These include *Tom Swift and His Jetmarine; Tom Swift and His Electronic Hydrolung*; *Tom Swift and His Triphibian Atomicar*; *Tom Swift and His Megascope Space Prober*; *Tom Swift and His Repelatron Skyway*; *Tom Swift and His Aquatomic Tracker*; *Tom Swift and His 3-D Telejector*; *Tom Swift and His Polar-Ray Dynasphere*; *Tom Swift and His Subocean Geotron*; *Tom Swift and His G-Force Inverter*; *Tom Swift and His Dyna-4 Capsule*; *Tom Swift and His Cosmotron Express*. As far as I know, Tom Swift never actually invented a time machine, but I have attempted to rectify that omission in the present chapter.

Tom and attempts to convince him that he, the older Tom, is the person the younger Tom will become in 10 years. The younger Tom is initially unwilling to accept this. But his scepticism dissolves once he grasps that the appearance of his older self automatically confirms that his efforts to construct a time machine will ultimately be crowned with success.

But *is* the younger Tom encountered by the time-travelling older Tom truly the latter's previous self? There is good reason to doubt it. Let us call Tom 1 the Tom at 20 who spends the following 10 years developing the time machine, and Tom 2 the Tom at 30 who has succeeded in building and testing the machine. Finally call Tom 3 the 20-year-old fellow who is encountered by Tom 2, ten years in the latter's past. Now, Tom 3 *cannot* be the same person as Tom 1. For, at 4 p.m. on Dec. 6, 1994, Tom 1 is sitting at his desk and, we shall suppose, sees nothing unusual. But *at that same time and place* Tom 3 is startled by the sudden appearance of Tom 2. Since Tom 1 and Tom 3 have had *different* experiences at the same time and place, then by the law of noncontradiction they *cannot be the same person.* Another way of drawing this conclusion is that Tom 1 slightly later has no memory of encountering himself, but Tom 3 presumably does. Accordingly the "Tom" Tom 2 encounters *after* his trip into the past (Tom 3) *cannot be* his past self (Tom 1). Tom 3 must then be, so to speak, a copy—a *temporal clone*—of Tom 1 induced by Tom 2's trip in to the past.

Another puzzle arises. When Tom 2 returns to the past and meets Tom 3, what in the meanwhile has happened to Tom 1? He was originally present at 4 p.m. on Dec. 6, 1994, but now seems mysteriously to have disappeared! It cannot be the case that Tom 1 has "become" Tom 3, because, as we have seen, the two cannot be identical. But at 4 p.m. on Dec. 6, 1994, it would seem that *both* Tom 1 and Tom 3 are sitting at the same desk. For this to make any kind of sense one can only conclude that there must also be a duplication or cloning of the *desks*, with Tom 1 sitting at one desk and Tom 3 at the other. But the two distinct desks would then occupy the same position at the same time. While this is logically possible, a more intelligible, if still radical, conclusion is that the two desks occupy *different realms of existence*, in short, different *timelines:* a timeline 1 containing the original desk with Tom 1 sitting at it; and a timeline 2 containing, along with Tom 2, the other desk at which Tom 3 sits. Thus we are led to conclude that Tom's travel into the past induces the original timeline 1 to *split* or *branch*, giving rise at 4 p.m. on Dec. 6, 1994—the time at which Tom 2 arrives in the past—to a new "branch"—what we have called timeline 2. Timeline 2 contains *two* copies of Tom at 4 p.m. on Dec. 6, 1994, namely Tom 2 and Tom 3.

Clearly, Tom 1 is present only in timeline 1, Tom 3 only in timeline 2, while just Tom 2 appears in both timelines and is the sole "Tom" directly aware of the splitting. At 4 p.m. on Dec. 6, 2004, Tom 2 "disappears" from timeline 1 to "reappear" in timeline 2 at 4 p.m. on Dec. 6, 1994 where he meets Tom 3. Meanwhile, timeline 1 continues to evolve past 4 p.m. on Dec. 6, 2004, but it no longer contains any version of Tom. As for timeline 2, "before" 4 p.m. on Dec. 6, 1994 it coincides with timeline 1, and contains just a single Tom, that is, Tom 1. At 4 p.m. on Dec. 6, 1994, timeline 2, containing Tom 2 and Tom 3, "branches off" from timeline 1. In this process, each object in timeline 2 is produced by "branching off" from the corresponding object in timeline 1. In particular Tom 3 is produced by "branching off" Tom 1. Before 4 p.m. on Dec. 6, 1994, Tom 3 can be thought of as "coinciding" with Tom 1, while after that time they differ and indeed inhabit different timelines. In timeline 1 Tom 1 continues to evolve into Tom 2 who later travels not just back in time, but into timeline 2 where he meets Tom 3.

Notice also that the appearance of three Toms is somewhat deceptive. Actually there are in fact just two Toms, corresponding to the two timelines. Tom 2 is an evolved version of Tom 1—a name for Tom 1 at a later time. Only the temporal clone Tom 3 is "new."

Two further, if macabre ways of demonstrating the non-identity of Toms 1 and 3 are suggested by the following variants of our scenario.

In the first variant, suppose that Tom 2, after some pondering, concludes that Tom 1 and Tom 3 cannot be the same person and decides to put this to the test by killing Tom 3. If Toms 1 and 3 were actually identical, reasons Tom 2, then the effect of killing Tom 3 would be the same as eliminating Tom 1 before he has evolved into Tom 2. In this case Tom 2 could not have come into existence, and so would "disappear." On the other hand if Toms 1 and 3 are different, then the elimination of Tom 3 should not have any effect (apart, possibly, from causing remorse) on Tom 2. Tom 2 duly kills Tom 3, but Tom 2, as anticipated, is unaffected. In this variant, Tom 3 dies in timeline 2 but Tom 2 remains very much alive there.

In the second variant, Tom 2 undergoes a mental aberration some time after his arrival in timeline 2 which causes him to construct a doomsday device which he threatens to set off. In a desperate measure, Tom 3 attempts to eliminate Tom 2 by committing suicide, reasoning that, if he (Tom 3) coincides with Tom 1, then such action will have the effect of eliminating Tom 1 before he has time to evolve into Tom 2, in which case Tom 2 could not have come into existence, and so would "disappear." (This gimmick appears at the end of the recent time-travel movie *Looper*.) Sadly, however, since Tom 3 is *not* Tom 1, the former's suicide succeeds only in removing *him* from the scene; the act has no effect whatsoever on Tom 2, who continues, in timeline 2, to pursue his diabolical scheme.

Another way in which Tom 3 might try to frustrate Tom 2's intentions would be for him to travel into the past and attempt to eliminate Tom 2 just as he steps out of the time machine at 4 p.m. on Dec. 6, 1994. The effect of this would be to induce a branching of timeline 2 into a new timeline 3 into which Tom 3 travels. The Tom who steps out of the time machine and whom Tom 3 attempts to kill is in fact not the original Tom 2, but a clone, Tom 4, in timeline 3. While Tom 3 can have no interaction with Tom 2 by travelling into the past, he could succeed in killing Tom 4, in which case any evil scheme planned by the latter would indeed be prevented. But this would take place in timeline 3, *not* as Tom 3 intended, in timeline 2. Events in timeline 2 remain unchanged.

Also note that if Tom 3 fails to kill Tom 4, then the two of them could then go on to meet the counterpart—Tom 5—of Tom 3 in timeline 3. This timeline contains *three* copies of Tom—Tom 3, Tom 4, and Tom 5. (This scenario is the basis of William Tenn's amusing short story *Me, Myself and I*.) Similarly, by means of successive trips into the "past" timelines can be generated which contain arbitrarily many copies of Tom—or indeed anything else. The opposition between the One and the Many could not be more vividly realized.

The *Grandfather Paradox* is a well-known example of a temporal paradox. Tom Swift's grandfather, a domineering, unpleasant man whom Tom detests, has made a fortune in business which has provided the funds for building Tom's time machine. Tom would like nothing better than to do away with him. Unfortunately, Tom's grandfather died of natural causes in 1990 while Tom was still an adolescent. But once Tom has built his time machine, he sees that his desire to kill his grandfather could be realized by travelling to the year 1940 when his grandfather was a young man, and killing him then.

Accordingly Tom purchases a rifle, becomes an expert marksman through target practice, travels in his time machine to 1940, conceals himself somewhere along the route of his grandfather's morning walk, and, rifle loaded, awaits his victim. His grandfather appears, Tom raises and aims his rifle, but what does he do next? There are two possibilities: (1) he pulls the trigger, or (2) he doesn't to pull it. If he pulls the trigger, so killing his grandfather a paradox ensues: his grandfather dies before Tom's father is born in 1950, so that Tom himself fails to be born in 1974. In that case, neither Tom nor his time machine ever came into existence, so that the whole sequence of events leading to the pulling of the trigger never occurred. In this case, the only logically consistent conclusion is that, when Tom pulls the trigger, the timeline splits into two branches, one the "original" timeline in which Tom's grandfather, his father, along with Tom himself pursue their lives as "before"; and the other, in which the grandfather dies in 1940, producing no descendants, but in which the "time-travelled" Tom suddenly appears in 1940. On the other hand, it is consistent to suppose that Tom does *not* pull the trigger, so that his grandfather is unaffected, goes on to sire Tom's father, who in turn sires Tom, who then constructs the time machine and travels into the past as before. In this case branching timelines are avoided so that the whole scenario can take place in a single timeline.

The American philosopher David Lewis has used the consistency of possibility (2), in which Tom's grandfather is not killed, to argue (in *The Paradoxes of Time Travel*, 1976) that backward time travel is possible within a single timeline. On Lewis's account, what we would say a time-traveller is capable of doing depends on what constraints exist for his or her actions (for further observations on this matter, see the section "Temporal Interdicts" below). Thus, in order for Tom to travel into the past and not produce a paradox, he must be in some way constrained so as not to pull the trigger in 1940. This constraint is imposed by facts about *future* events (viewed from 1940), for example, that Tom's grandfather made a business fortune enabling the time machine to be built, and, not least, that Tom himself came into existence. In order for time travel within a single timeline to take place, facts about the future must act as constraints, since, on logical grounds, no one can do anything incompatible with those facts.

While Lewis's account is consistent, it still leaves the puzzling question as to the *source* of the constraints, *from Tom's point of view*, preventing him from killing his grandfather. He could fail to kill his grandfather for a number of reasons: he relents at the last moment; he becomes aware that *if he did so*, he would cause a split in the timeline in which his father, of whom he is very fond, would not exist; the rifle jams; his grandfather is wearing a bulletproof vest, etc. The point is that Tom *could* kill his grandfather, but *doesn't*. Lewis is content to observe that the capacity to perform an action does not imply that the action will actually be carried out. This applies even if Tom decides to eliminate the "human factor" by sending a drone into the past programmed to kill his grandfather. In this case the element of conscious intention, or "changing one's mind" is not present, but still the drone could be shot down, or miss its target, etc.

Harold Ramis's 1993 movie *Groundhog Day* offers an especially engaging time travel scenario. The film opens on February 1, when Pittsburgh TV weatherman Phil Connors travels with his team to a small town in Pennsylvania to cover the annual Groundhog Day festivities on February 2nd. The following morning, he sends his report on the festivities and then attempts to travel with his team back to Pittsburgh, but is prevented from doing so by bad weather. The team is forced to return and stay another night in the small town. Phil wakes up the following morning on what he expects to be February 3rd, only to find, to his bewilderment, that he is reliving February 2nd. The day wears on in exactly the way as he recalls it, yet nobody but he seems to be aware of the repetition. At the end of the day he falls asleep and, on awakening, again proceeds to relive February 2nd. After undergoing a number of similar reawakenings on February 2nd, he realizes that, apart from affecting his own memory, there is no trace in the outer world, or in other people's memories, of the actions he recalls taking the previous "day." He becomes increasingly desperate to escape the repetition and resorts to suicide, only to find that he again wakes up, fully restored, the following day. Eventually he decides to use the repeated days to improve himself by developing new skills such as mastering the piano and learning to speak French. The movie ends with his finally waking up on February 3rd, and continuing with a normal life, yet still possessing the skills he has acquired through his repeated trips into the past.

Is this scenario consistent with our "branching timelines" account of travel into the past? Yes, with some minor changes. In the "branching timelines" setup, Phil would be an involuntary time traveller repeatedly thrust back one day into the past, but each time this happens he "returns," not to the timeline—call it timeline *A*—in which he fell asleep the night "before," but to a *new* timeline *B* branching off from *A* on the morning of February 2nd. (This has to occur in such a way that Phil does not encounter his previous self.) The original Phil's timeline continues to evolve in universe *B*, which coincides with timeline *A* up to the instant at which the split occurs. Apart from Phil himself, everything in timeline *B* is a clone of its counterpart in timeline *A*. Thus Phil is the only person in timeline *B* who is actually aware that a splitting of the timeline has taken place.

In Ramis's movie Phil undergoes this cyclic process of "returning" to the morning of February 2nd repeatedly. In the "branching timeline" formulation Phil's continued existence *as the original Phil* is required to enable him to pursue his repeated hops into the branching timelines induced by the temporal cycles. He is the one constant element within a series of timelines undergoing continual duplication. This is not compatible with his committing suicide, as takes place with dramatic effect in the film, since then he would not be able to repeat his trip into the past, as the film requires. Also, during each temporal cycle Phil becomes a day older. Clearly the number of times that he can return to the past is then bounded by his life span. If we agree that he cannot live more than 100 additional years say, then he cannot undergo more than 100 × 365 = 36500 temporal cycles. Moreover, each time Phil materializes in timeline *B*, he will *appear* to be a day older than he was in timeline *A*. This aging effect is cumulative, and would eventually be noticed by his acquaintances in universe *A*. To avoid this, Phil would probably have to age no more than 10 years, i.e., undergo no more than 3650 temporal cycles.

Gregory Benford's memorable novel *Timescape* (1980) contains an example of a "split" timeline. In Benford's scenario, the world of 1998 is a growing nightmare of desperation, of uncontrollable pollution and increasing social unrest. Two scientists in Cambridge are conducting experiments with tachyons—subatomic particles that travel faster than light and therefore, according to the theory of relativity, can move backwards in time (see Chapter V). The scientists attempt to use tachyons to send a message into the past warning the previous generation of the dire situation in their future, so that they, forewarned, may take steps to prevent it. In 1962, a young California scientist finds his experiments are being spoiled by unknown interference. As he begins to suspect something near the truth, it becomes a race against the clock—the world is collapsing and will only be saved if he can decipher the messages in time. But acting on the messages from 1998 he succeeds in deciphering has the effect of splitting the original timeline into two different timelines—one in which the ecocatastrophe is avoided (and in which President Kennedy is shot, but not killed), and the "original" one from which the messages were initially sent, in which the ecocatastrophe cannot be prevented.

A colourful take on travel into the past is offered by Alfred Bester in his clever story *The Men Who Murdered Mohammed* (1958). In Bester's story, Henry Hassel, Professor of Applied Compulsion at Unknown University, returns home one afternoon to find his wife in the embrace of one of his colleagues. Furious, he throws together a time machine, travels into the past, and shoots his wife's grandfather. He returns to the present confidently expecting that his wife will no longer exist, but instead finds her in exactly the position in which he left her. Quickly inferring that faithlessness must run in his wife's family, he again jumps into his time machine, returns to the past and kills his wife's maternal grandmother. But he returns to the present a second time to find his wife in the same position. Astounded, he makes increasingly desperate efforts to affect the present by going into the past and slaughtering Columbus, Napoleon, and half a dozen other celebrities, all with no effect whatsoever. After a while he finds that nobody at all can see or hear him; he has in effect become a ghost. Later he meets another spectral time traveller who explains what has happened:

> Time is entirely subjective, a private matter. Time travellers travel into their own past, and into no other person's. There is no universal time. There are only billions of individuals, each with his own continuum; and one continuum cannot affect another. We're like millions of strands of spaghetti in the same pot. No time traveller can ever meet another time traveller in the past or future. Each of us must travel up and down his own strand alone. The fact that we're meeting each other now is explained by the fact that we're no longer time travellers—we've become part of the spaghetti sauce. You and I can visit any strand we like, because we've destroyed ourselves. When a man changes the past he affects only his own past—no one else's. The past is like memory. When you erase a man's memory, you wipe him out, but you don't

wipe out anyone else's. You and I have erased our past. The individual worlds of the others go on, but we have ceased to exist. With each act of destruction we dissolved a little. We've committed chronicide.

In this story Bester shows brilliantly how time embodies the opposition between the Subjective and the Objective.

H.G. Wells's *The Man Who Could Work Miracles* (1899) offers a logically impossible, but highly entertaining trip into the past. In the story, a Mr. Fotheringay finds that he has in some mysterious way acquired the power of having his every wish miraculously granted. Beginning modestly with a few minor miracles, such as transforming his tobacco-jar into a bowl of violets, one night Mr. Fotheringay is encouraged by the local parson, Mr. Maydig, to perform bigger feats. So he duly drains a swamp, improves the railway, and, in a reforming zeal, transforms all the alcoholic beverages in the vicinity to water. It grows late, and Mr. Fotheringay (evidently a man of little imagination) starts to worry about getting to work the following day. Mr. Maydig suggests that Mr. Fotheringay, like a latter-day Joshua, stop the hour growing any later by arresting the rotation of the earth. Mr. Fotheringay obliges, but as a result finds himself pitched abruptly into a whirling chaos; for in stopping the earth from rotating he has neglected to arrest the motion of the objects on its surface, which have as a result all been thrown violently forward at high speed. Failing to understand this, and thinking that his miraculous powers have gone wrong, Mr. Fotheringay wishes that he be rid of them, and that time run back to the moment immediately prior to their appearance. Thus everything returns to normal, and so in actual fact Mr. Fotheringay never possessed the power to work miracles at all.

TEMPORAL LOOPS

Another strange, but apparently possible result of travel into the past is the production of *closed temporal loops*. Let us return to our Tom Swift scenario, but now suppose that, on 4 p.m. on Dec. 6, 1994, Tom 1 *actually witnesses* the arrival of the time machine containing his older self (Tom 2), so that Tom 1 and Tom 3 are *identical*. In that case there is no splitting of timelines. And since Tom's arrival in the past has *already happened*, this fact cannot, on pain of logical contradiction, be changed. Therefore Tom *must* travel into the past to meet his younger self. Tom is, so to speak, "trapped" in a closed temporal loop in which he is *foreordained* to build the time machine and, 10 years later, travel back in time. This can be seen as an extreme instance of the opposition between Chance and Necessity, since in the temporal loop chance has been *entirely replaced* by necessity, as far as Tom's actions are concerned.

After 4 p.m. on Dec. 6th, 1994, the single timeline contains *both* Tom 1 and Tom 2, Meanwhile the Tom 2 who has arrived in the past continues to age and remains permanently 10 years older than Tom 1. One can imagine the older Tom 2 giving the older Tom 1 an affectionate farewell as the latter steps into the time machine to close the temporal loop.

Now let's suppose that, before Tom 2 can step out of the machine, a malfunction causes the machine to explode, destroying it and killing Tom 2. As in the previous scenario, if Tom 2 is *truly* Tom 1's future self, then in travelling to the past Tom 2 has *not* induced a past split in the timeline; that is, the whole affair is confined to the original timeline 1. So, if Tom 2 really *is* Tom 1's future self, then again, on pain of logical contradiction Tom 1 *will have constructed and used* the time machine, only this time *he causes both his own future death and his witnessing of it*. Here again Tom 1 *has no choice* in the matter, despite his awareness of the consequences. Tom 1 is thus a "fatalist" trapped in a closed temporal loop, in which he sees himself—like Sisyphus endlessly rolling the rock uphill—repeating the same actions that he knows will lead to his death. This scenario is the basis for the plots of Chris Marker's film *La Jetée*, Terry Gilliam's film *12 Monkeys*, and Philip K. Dick's story *A Little Something for Us Tempunauts*.

In this "fatalist" scenario Tom is trapped in a 10-year temporal loop in timeline 1 between Dec. 6th, 1994 and Dec. 6th, 2004. But Tom is the sole occupant of the timeline to be thus trapped. After Tom 2 enters the time machine at 4 p.m. on Dec. 6th, 2004, the other occupants of timeline 1 continue on unaffected into that timeline's future—from which Tom 2 has disappeared!

On the other hand, suppose that Tom 1 decides at some point to exercise his free will and not to build the time machine, so avoiding the doom he has foreseen. In that case, as science fiction writers like to say, the closed temporal loop is "broken." What this means is that, at the point Tom 1 takes the decision to exercise his free will, he "splits"—along with the timeline—into two: the original "fatalistic" Tom and a different "free" version in which he takes his future into his own hands. For this second Tom the future is open; for the first, trapped in a temporal loop, it is closed. Both can continue to exist without inconsistency in their now distinct timelines. But the "fatalistic" Tom must, necessarily, either have abandoned, or have been deprived of, his free will. In effect, he has become a zombie.

Another odd but possible by-product of travel into the past is the creation of a *closed causal loop*. By this is meant circumstances involving events each of which is the cause of the other.

Consider, for example, the *Oedipus* myth. In the best-known version of the myth—the one dramatized by the playwright Sophocles—Oedipus grows up to murder his father Laius and marry his mother Jocasta. A time-travel variant could have Oedipus growing up, travelling into the past, marrying his mother and siring himself. Thus Oedipus, far from murdering his father, actually *is his own father*. This scenario could in principle be realized within a single timeline. In that timeline, the adult Oedipus's trip into the past could take place *without changing that past* and so inducing no splitting of the timeline.

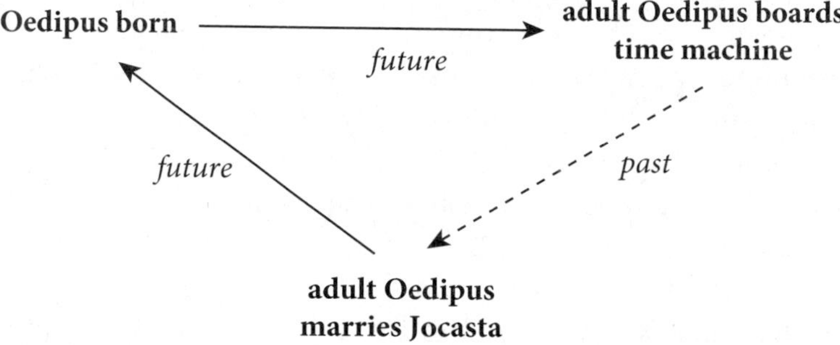

The closed causal loop is—like actually occurring ordinary events—a curious but *unalterable* feature of the timeline. Oedipus's birth is the cause of the existence of the adult Oedipus, which in turn is (part of) the cause of Oedipus's birth.

Robert Heinlein's story *All You Zombies* provides the ultimate example of a closed causal loop of this type. By travelling backwards and forwards in time, and undergoing a sex-change operation, the story's central character creates a closed causal loop in which he becomes *both* of his parents. This story provides the basis for the 2014 movie *Predestination*.

Here is a story with a different example. Early in the year 1599 William Shakespeare is suffering from writer's block. The Bard is standing at his desk one day, quill in hand, but devoid of inspiration. Suddenly a hole opens up in the wall of his room and a man steps through it. He bears a manuscript which he hands to the astonished Shakespeare. It is entitled *The Tragedy of Hamlet, Prince of Denmark*, by William Shakespeare. The man explains that he is a professor of literature from several hundred years in the future, and that he has travelled back in time to 1599. In his time, he says, *Hamlet* is one of the most celebrated works of literature, and he has admired it since he first read it in his youth. The professor then takes his leave and re-enters the hole in the wall; it closes, and he vanishes, leaving the manuscript in Shakespeare's hands. After the professor's departure, Shakespeare reads through the manuscript, finds it intriguing, and decides to pass it off as his own work. He never reveals its true source.

But what, precisely, is *Hamlet's* true source? There seem to be (at least) two possibilities, one in which the timeline splits, the other involving a closed causal loop.

In the first of these possibilities, Shakespeare himself actually wrote *Hamlet* sometime in 1599 (or soon thereafter), and the professor materializes early in 1599 *before* the play was written. In this case, as with the Tom Swifts, there must be *two* Shakespeares, the "original" one who actually wrote *Hamlet* and the other a temporal clone whom the professor actually meets. There must also be a corresponding splitting of timelines into two branches, the first—the "original" timeline—containing the original Shakespeare, and the second containing the temporal clone. (Note that, up to the point of the professor's arrival, the second timeline coincides with the original one.) In this case *Hamlet* does not get entangled in a causal loop. The source of *Hamlet* is the original Shakespeare. This remains the case in the second timeline, even though the original Shakespeare is no longer present in the second timeline after the professor's arrival.

The second possibility pivots on setting the whole scenario within a *single* timeline. In this case a causal loop is generated. Here we suppose that *Hamlet* was *not* written by Shakespeare through his own inspiration (in any timeline), but that he does actually receive the manuscript of *Hamlet* from the time-travelling professor in 1599, and passes it off as his own work. This scenario could, in principle at least, consistently take place without requiring the timeline to split. But then, while avoiding logical contradiction, we are left with the sheer oddness of the resulting closed causal loop. For let us ask the question: who actually wrote *Hamlet*? It cannot be the professor, since he first read the play in his youth. It cannot be Shakespeare himself, since he plagiarized it. Then who? Nobody, it would seem. *Hamlet* has materialized causelessly as the centre of a loop in time.

An ingenious variation on this idea was presented by William Tenn in his science fiction story *The Discovery of Morniel Mathaway*. A professor of art history from the future time travels some centuries into the past in search of an artist whose works are celebrated in the professor's time. On meeting the artist in the flesh, the professor is dismayed to find the artist's current paintings talentlessly amateurish. The professor happens to have brought with him from the future a catalogue containing reproductions of the paintings later attributed to the artist. He has come to see that they are far too accomplished to be the artist's work. He shows the catalogue to the artist, who has quickly grasped that in the future he will be welcomed as a celebrity. By means of a ruse, the artist succeeds in using the time machine to travel into the future, taking the catalogue with him. This strands the professor in the "present." To avoid entanglements with authority the professor assumes the artist's identity and later achieves fame for producing what he believes are just copies of the paintings he recalls from the catalogue. This means that *he*, and not the artist, created the paintings in the catalogue. But he could not have done so without having seen the catalogue in the first place, and so we are presented with a causal loop.

Philip K. Dick offers a further variation on this theme in his short story *Orpheus With Clay Feet*. Here the principal character, Jesse Slade, employs a time travel agency to arrange a trip into the past so as to enable him to act as a muse for some significant historical figure. Slade chooses to stimulate his favourite science fiction writer, the great Jack Dowland (which is also the pseudonym Dick used for this story). But in his efforts to inspire Dowland, Slade divulges that he is a time traveller hoping to inspire his work. Dowland is highly offended by this, and as a result, never becomes the major science fiction writer he was meant to be. He does, however, publish, under the pen name Philip K. Dick, a single science fiction short story called *Orpheus With Clay Feet*, about a time traveller who travels into the past to inspire his favourite science fiction writer, one Jack Dowland.

TIME TRAVEL INTO THE FUTURE

Let us now turn to *time travel into the future.* While seemingly less problematic than travel into the past, difficulties still arise. Let us return to the exploits of Tom Swift, and suppose that in timeline 2, Tom 2 constructs the doomsday device so that it cannot be prevented from going off at a certain instant, but after doing so comes to his senses. He then travels into the future in an attempt to prevent himself from using the time machine[2] to travel into the past and build the doomsday device. Could he succeed? Apparently not, because a successful outcome would require him to travel to timeline 1 to prevent the Tom 2 in *that* timeline from travelling to timeline 2 and setting up the doomsday device. But it is logically impossible for the Tom 2 in timeline 2 to do this, since the Tom 2 in timeline 1 did *in fact* travel to timeline 2 and later set up the doomsday device. He cannot both travel and not travel to timeline 2. Tom 2 *could* travel into the future while remaining in timeline 2, but he would not encounter himself there.

We may also ask if it is possible, in timeline 1, for an "intermediate" Tom—Tom 1.5, say—temporally located between Toms 1 and 2 and in possession of the time machine, to travel into the future and prevent his future self Tom 2 from travelling to timeline 2. Now this, again is logically impossible if taken literally. For if Tom 2 actually does travel to timeline 2 at a certain moment, he cannot be prevented from doing so, by Tom 1.5 or anybody else. Tom 2 cannot, at a certain time, both travel and not travel to timeline 2. On the other hand, it is logically possible for a temporal clone of Tom 1.5—call him Thomas—to travel into the future of another universe cloned from universe 1, call it universe 1.5, in which he meets a counterpart of Tom 2, Tom 2.5 say, whom he prevents from travelling into the past. If timeline 2 was produced by a "past" split, timeline 1.5 may be thought of as the result of a "future" split. While timeline 1 contains Toms 1, 1.5, and 2, timeline 1.5 contains Toms 1 and 2.5, along with Thomas. In timeline 1, Tom 1 evolves into Tom 1.5, who in turn evolves into Tom 2, while in timeline 1.5, Tom 1 evolves into Thomas, who in turn evolves into Tom 2.5.

---

2 The unwritten book, *Tom Swift and His Time Machine*, would include the line "'*Get out of that time machine fast,' said Tom swiftly.*"

THE FUTURE TIME VIEWER

A related theme is that of the *future time viewer*. Let us suppose that Tom Swift has, after a number of failed attempts at constructing an authentic time machine, decided to lower his sights and design a *time viewer*—an instrument which, like a crystal ball, allows any event in future time to be observed. He succeeds in this endeavour and, eager to try his device out, proceeds to scan the near future. He first scans the stock market and learns that the value of certain stocks will soar in the coming months. He invests heavily in these, making a handsome profit. The Tom who receives the money is the same Tom (if a little older) who invests on the basis of the information about the future furnished by the time viewer. So the time viewer could, it would seem, be used consistently in a single universe.

But a little later Tom is shaken when his scanning of the future reveals that, on the following day, a terrorist will detonate a nuclear bomb where he lives, in New York City, killing millions of people, including himself. The time viewer enables him to see exactly where and when the device will be placed. He immediately contacts the authorities and informs them of the bomb's location. They defuse the bomb and millions of lives—including Tom's—are saved.

Now here the question arises: *exactly whose lives are saved* by Tom's actions? Assuming that the time viewer truly reveals the future, then the events it reveals, including the deaths of millions of New Yorkers, will *actually* take place, and so cannot be "changed." On the other hand, by informing the authorities Tom has apparently prevented the deaths of millions. Now the "millions of New Yorkers" cannot both die and not die, so there must be *two* groups of New Yorkers, one group which dies and another which survives. In that case, Tom's actions must have caused the timeline to split into two, the original one, timeline 1, in which millions of New Yorkers, including a doomed Tom, die, and a second one, timeline 2, in which the millions, including a saved Tom, live on.

Notice that, if the time viewer truly reveals the future, then as far as the denizens of timeline 1 are concerned, *Tom's actions are irrelevant*. The doomed New Yorkers of timeline 1 will die no matter what Tom does. If he does nothing, this is clear. Yet if he informs the authorities, then timeline 2 branches off and the New Yorkers of *that* timeline are saved. While this is surely a good thing, Tom's action, sadly, it would seem, has no effect whatsoever on the doomed New Yorkers of timeline 1.

Objectively, this analysis appears logically correct. But *subjectively*, profound difficulties remain. Consider the whole scenario from Tom's point of view. He is sitting at his time viewer, he sees the bomb explode and he himself and millions of others die. Through his mind the thought runs: I and millions of others will die, but thanks to the time viewer I am aware in advance that this is going to happen. Since I accept the evidence of the time viewer, I realize that I can do nothing to alter the events shown on it, but I can, by informing the authorities of the location of the bomb, cause the timeline to split in such a way as to ensure that the bomb does not explode in the branch timeline (timeline 2) thereby created. Tom is now confronted with a choice. He could do nothing, in which case if he rechecks the time viewer, he will again see the explosion and the deaths, confirming that he is the doomed Tom and still occupies timeline 1. On the other hand he could contact the authorities, in which case (assuming that the authorities act to defuse the bomb), on rechecking the time viewer he will see no explosion and no deaths, so confirming that he must be the saved Tom in timeline 2. The New Yorkers of timeline 1 continue on to their doom, but only Tom 1 is aware of their (and his) fate in advance. The New Yorkers of timeline 2, including the saved Tom, continue on to their salvation, but only the saved Tom is actually *aware* that they are the result of a split in the timeline. But Tom *knows* that a split in the timeline must occur when he takes action. In that case, immediately after he takes that action, which Tom is *he*? The doomed, or the saved? Had Tom taken no action, only the doomed Tom would exist. By acting, Tom creates a new version of himself—the saved Tom. We are familiar with physical things splitting into two, and can accept in principle that they could even be duplicated. But it is extremely difficult to make sense of the idea that an individual consciousness can be so split. Tom is fully aware of the fact that, by taking the action he does, he is thereby creating a duplicate of himself. So he will ask himself, immediately after taking the action, *which Tom am I?* To compound the difficulty, *both* Toms will ask themselves the question. One is saved, one is doomed. But the doomed Tom could well ask himself, why am *I* the doomed Tom? Of course, objectively he can answer this question by noting that he, the doomed Tom, was the one who failed to act. But from a subjective point of view he can ask: why was *I* the Tom who failed to act? Why couldn't *I* have been the saved Tom? There seems to be no satisfactory answer to this question.[3]

There is a further difficulty, which seems to undermine the whole concept of a future time viewer. For what if Tom views the future *before* the moment he sees the explosion, informs the authorities and so splits the timeline. *Which future does he see?* Is it the future seen by the doomed Tom, or the one seen by the saved Tom? This is far from clear, because at the earlier time, before the timeline splits, there are *two* actual ways in which the timeline will in fact develop in the future, one in which the explosion occurs, and the other in which it is prevented. At that earlier time, the viewer would have to display *both* timelines since they are *both* part of the future. The viewer cannot "choose" between these two futures. Moreover, there are many more possible futures, for example one in which Tom fails to contact the authorities but the bomb fails to go off, another in which Tom contacts the authorities but they fail to arrive on time to defuse it, in endless variation. The time viewer would, in principle at least, have to display all of these possibilities, thereby rendering it effectively useless. For in the final analysis it would have to display *all* logically possible futures without indicating *which* is the "actual" one. In fact it would appear that the concept of an "actual" future has been rendered meaningless. This seems to be a necessary consequence of the presence of future branching timelines.

The existence of future branching timelines also makes the whole concept of travel into the future problematic. For the question arises, to *which* future timeline does a time-traveller actually travel? Suppose that Tom has constructed two time machines and he invites his friend Jerry to join him in a trip of exactly one year into the future, but in *separate* time machines. When Tom and Jerry step out of their respective time machines one year in the future, what guarantee do they have that they will step out into the *same* timeline? Since during the actual year that has passed many branchings of the original timeline may have occurred, no such guarantee seems possible.

---

3 This echoes the haunting question Hermann Weyl imagines the betrayer of Christ asking: *Why did* I *have to be Judas*? (*Philosophy of Mathematics and Natural Science*, 125).

TWO-DIMENSIONAL TIME

The American philosopher Jack Meiland has attempted to resolve some of the paradoxes of time travel by arguing that time could have *more than one dimension*.[4] His basic idea is that the past can change from moment to moment. This is explicated—for two dimensional time—by the following diagram:

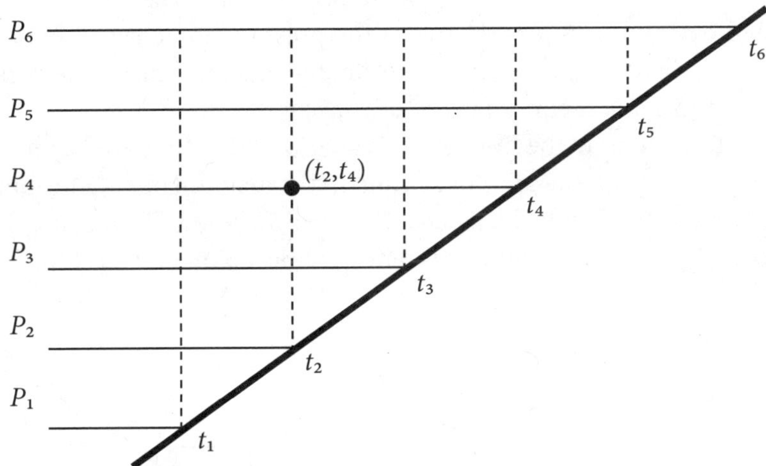

Here the bold diagonal line represents the evolving present—the timeline—of a given conscious observer and the points t_i on it particular present moments in his life. In the ordinary way, the conscious observer's timeline is *one-dimensional* and at each moment t_i his past would be represented by the portion of the bold diagonal line "before" t_i. Thus, on this timeline, the past "grows" continuously from moment to moment in such a way that the past at a present moment is always an extension of the past at any previous moment. Let us call this property the *linearity condition* on the past.

Meiland proposes to *drop* the linearity condition and allow the past to vary with the present moment in a much freer way. For example, consider two moments t_1 and t_2 with t_1 earlier than t_2, and an event E_1 occurring at t_1. Supposing the past to be linear, at the moment t_2, the event E_1 must still have occurred at the past moment t_1. But in Meiland's account this is not necessarily the case: while the event E_1

4 "A two-dimensional passage model of time for time travel," *Philosophical Studies* 26 (1974): 153–73.

occurred at t_1 *when* t_1 *was the present moment*, it need no longer have occurred at t_1 *when* t_2 *is the present moment.*

Meiland's temporal model thus posits a radical, nonlinear evolution of the past with the present moment. This is represented in the diagram above by the horizontal lines P_1, ..., P_6. Each P_i represents the "past" at the present moment t_i. The vertical dotted lines represent the "projections" of previous moments into the evolving past. Thus, for example, the point (t_2, t_4) is the moment t_2 in the past P_4, or, if you like, the past moment t_2 *relative to* the present moment t_4. Moments are accordingly assigned *two* coordinates, making the framework a *two-dimensional* theory of time.[5]

Let us see how the Tom Swift scenario with which we began our discussion is presented in this setting. In the diagram below, the longer bold diagonal line may be taken to be the time line of Tom 2, the time t_1 to be 4 p.m. on Dec. 6th, 1994, and t_4 to be 4 p.m. on Dec. 6th, 2004. At the moment t_1 Tom 2 coincides with Tom 1. At t_4 Tom 2 boards his

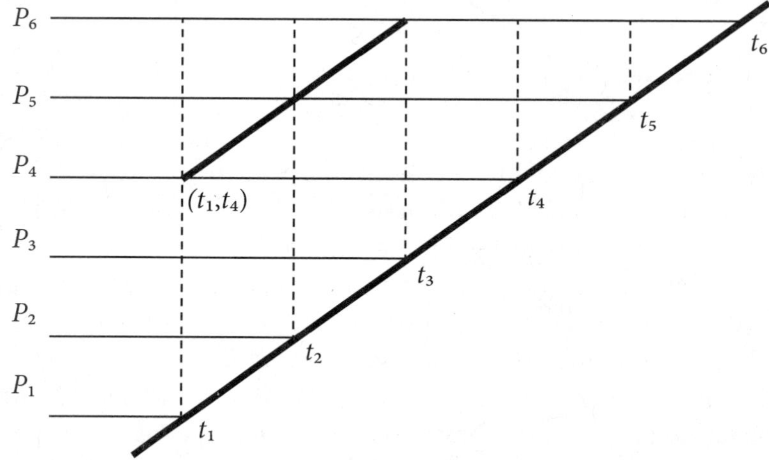

time machine, travels back to the point (t_1,t_4) in the evolved past P_4, where he meets Tom 3, the counterpart of Tom 1 in P_4. Thus the "split" in the timeline caused in our original scenario by Tom 2's travel into the past is avoided in Meiland's setting, being effectively replaced by travel into an evolved past.

---

5 Notice that then, strictly speaking, each moment t_i on the bold diagonal line should be represented by the point (t_i, t_i).

Once Tom 2 arrives at (t_1,t_4) he can continue to remain in the past, and his timeline from that point is represented in the above diagram by the shorter bold diagonal line. His total timeline is accordingly represented by the portion of the longer bold diagonal line stretching from t_1 to t_4 *together with the shorter bold diagonal line*. That being the case, the whole of the longer bold diagonal line would have to be taken to represent the timeline of an observer *who does not undertake time travel*.

The reader may find it entertaining to try to represent Tom Swift's further temporal exploits in Meiland's model.

Philip K. Dick's 1954 novella *A World of Talent* postulates what amounts to a two-dimensional temporal universe, in which one of the characters sees time as a chessboard and possesses the paranormal ability to "select" the past.

TEMPORAL INTERDICTS

Some students of time travel balk at the idea of branching timelines and their complexities. They hold that logically impossible trips into the past would simply be blocked on physical grounds alone. The great logician Kurt Gödel made the intriguing suggestion that such temporal journeys might be prevented by limitative principles of physics such as the uncertainty principle of quantum mechanics that a particle's position and momentum cannot be determined simultaneously (discussed in Chapter VI) or the postulate of special relativity that nothing can travel faster than light (discussed in Chapter V).

There is, however, an important difference between limitative principles of physics such as these and any principles (call them "temporal interdicts") invoked to block changes of the past. In the case of the uncertainty principle it is *logically possible* that it could be violated, so allowing an electron's position and momentum to be simultaneously measured with pinpoint precision. In the case of the light postulate, it is *logically possible* for a body's velocity to exceed that of light. But, as has already been shown, any violation of a temporal interdict would involve a *logical contradiction*. Unless what we take to be objective reality is merely an illusion, or the law of non-contradiction can be violated, there can be no infractions of temporal interdicts.

We have seen that closed causal loops are consistent and accordingly not excluded as possible outcomes of trips into the actual past.

But how could any temporal interdict devised expressly to prevent time travel for the purpose of *changing the past* not at the same time also frustrate time travel for the purpose of *setting up closed causal loops*? For example, imagine this change in William Tenn's story "The Discovery...": the professor, insanely jealous of the artist's fame, resolves to travel a little further into the past with the intent of suffocating the artist as an infant in his cradle. This would have to be impossible if, as stipulated in the story, the artist *in fact* lived to adulthood. In that event, the professor's evil design *must* be frustrated on pain of logical contradiction. But how? By the professor failing to complete his journey? If the professor's trip into the past could actually be completed in the original nonparadoxical case, it could surely also be completed in the second, seemingly paradoxical case. How could the time machine itself distinguish between its operator's intentions in the two cases? So what then remains to prevent the professor, upon arriving at his temporal destination, from suffocating the infant, thereby creating a contradiction? Nothing, it would seem, apart from contrived coincidences such as the professor suddenly dropping dead, or the infant's parents appearing and getting the professor arrested.

If the professor *does* succeed in suffocating the infant, then the "past" into which he has travelled is in fact a different "past" from the one in which he originated. As in the Tom Swift case, the professor's actions have caused the timeline to split into two distinct past branches: one in which the artist survives into adulthood, and another in which he dies in infancy. The professor's efforts to transform the unique "actual" past, in which the artist truly existed, into one in which he ceases to exist, is doomed to futility on purely logical grounds.

TIME TRAVEL AS A PHYSICAL POSSIBILITY

So far our discussion of time travel has been purely speculative. But the idea of travel into the past has arisen as a serious possibility within theoretical physics itself. In 1949 Gödel constructed the first mathematical models of the universe in which *travel into the past is, in theory at least, possible*. Within the framework of Einstein's general theory of relativity Gödel produced models of the cosmos which contain so-called *closed time-like curves*, that is, curves in spacetime which, despite being closed, still represent possible paths of bodies. An object moving along such a path would travel back into its own past, to the very moment at which it "began" the journey. More generally, Gödel showed that, in his "universe," for any two points *P* and *Q* on a body's track through spacetime (its *world line*), such that *P* temporally precedes *Q*, there is a timelike curve—that is, a curve representing a path the body might actually follow—linking *P* and *Q* on which *Q* temporally precedes *P*. This means that, in principle at least, one could board a "time machine" and travel to any point of the past.

From this Gödel drew the remarkable inference that, in such a situation, time or, more generally, change, is not an objective phenomenon, but an *illusion* arising from our special mode of perception. (The idea that time is illusory has in fact been held by a number of philosophers, including Parmenides and Kant.) To see this, consider an observer initially at point *P* (with time coordinate t seconds as indicated by his own clock). At point *Q* (with time coordinate t') he boards a time machine and travels back to point *P*, taking time t'' to do so. In that case, according to his own clock, $t' - t + t'' > 0$ seconds have elapsed, and yet an identical clock left at *P* would show that 0 seconds have elapsed. In short, there has been no "objective" lapse of time at all.

Gödel also raises the issue of whether the fact that objective lapses of time fail to exist in his universe has any consequences for the universe in which *we* live—for us, at least, the actual universe. He points out that, while our universe differs observationally in certain respects from his model, there might be models containing closed timelike curves which are observationally indistinguishable from ours (a possibility later confirmed). In that case, it is already *possible* that our universe is one in which objective time is an illusion. And in any event, he goes on to say,

> The mere compatibility with the laws of nature of worlds in which there is no distinguished absolute time and in which, therefore, no objective lapse of time can exist, throws some light on the meaning of time also in those worlds in which an absolute can be defined. For, if someone asserts that this absolute time is lapsing, he accepts as a consequence that whether or not an objective lapse of time exists (i.e., whether or not a time in the ordinary sense of the word exists) depends on the particular way in which matter and its motion are arranged in the world.[6]

---

6 This is because in general relativity the geometry of the universe is determined by the distribution of matter in it.

> This is not a straightforward contradiction; nevertheless, a philosophical view leading to such consequences can hardly be considered as satisfactory.

Materialism is any philosophical view in which basic reality is nothing but matter and its motion. But it would be a curious materialism indeed which made the very existence of objective time depend upon the distribution of matter!

Branching timelines have arisen in connection with the so-called *many-worlds interpretation* of quantum theory (discussed in Chapter VI). In this interpretation, when certain types of interaction occur (typically, measurements), the timeline, or universe if you prefer, splits into different branches, one for each possible outcome of the interaction. Observers branch (or split) as well, and each observer on each branch sees one of the possible outcomes. Recent work by the physicist David Deutsch has shown that time travel into the past is compatible with the many-worlds interpretation. But again observe that here time travel takes place from the present of one "branch" of the universe into the "past" of *another* branch. There remains the problem of devising physically plausible "temporal interdicts" to block logically impossible trips into the past of a single universe, yet still permitting the logically possible ones.

Chapter V

Puzzles and Paradoxes of Relativity Theory

Einstein's theory of relativity exemplifies several oppositions: Absolute/Relative, Constant/Changing, Bounded/Unbounded, and in a certain sense, Finite/Infinite.

SPECIAL RELATIVITY

In 1905 Einstein formulated the *special theory of relativity*, a fundamental new theory of space and time. Its key postulate is that the speed of light (or any form of electromagnetic radiation) in a vacuum is the same for all observers, regardless of their motion relative to the light source. Einstein was led to formulate this startling hypothesis through his reflections on Maxwell's equations for electromagnetism, the well-established equations from the nineteenth century linking electricity, magnetism, and radiation. These predict that light has a specific, constant velocity in a vacuum, with the implication that the vacuum is fixed in Newtonian absolute space. Einstein saw that the validity of Maxwell's equations does not depend on the assumption of the presence of an absolute background space. He became convinced that Maxwell's equations should hold not only in a putative "absolute space," but in any *inertial frame*—a physical frame of reference resembling "absolute space" to the extent that Newton's first law of motion (the *law of inertia*) holds in it: any free motion of a body has a constant magnitude and direction. Einstein's postulate has the consequence that time and space are linked in such a way as to maintain the constancy of the speed of light regardless of the relative motions of sources and observers. If we call an *inertial observer* an observer occupying an inertial frame, then the reigning principle of Einstein's theory, the *principle of relativity*, can be put in the following way: *all physical laws should take the same basic form for all inertial observers.* Another way of putting this is: *all inertial observers are equivalent.*

For most everyday phenomena, in which velocities are much lower than the velocity of light, the predictions of special relativity are almost identical to those of classical mechanics. But it makes startlingly different predictions for objects moving at very high speeds. Remarkably, these predictions have all been experimentally confirmed.

Some of the most striking non-classical predictions of special relativity are:

- The *relativity of simultaneity*: Observers in a state of relative motion may disagree as to whether two events occur at the same time or whether one occurs before the other. This provides a nice instance of the Absolute/Relative opposition, since before Einstein it had been taken for granted that simultaneity of events was an absolute notion, independent of the observer.

- *Time dilation*: An observer watching two identical clocks, one moving and one at rest, will judge the moving clock to run more slowly. As the velocity of the moving clock approaches the speed of light, the observed rate at which the clock runs tends to zero.

- *Length contraction*: Along the direction of motion, a rod moving with respect to an observer will be measured as shorter than an identical rod at rest. As the velocity of the rod approaches the speed of light, the observed length of the rod tends to zero.

- *Relativistic mass increase*: The mass of a moving body increases with its speed, and tends to infinity as its speed approaches the speed of light.

- The *equivalence of mass and energy*: Mass and energy are convertible into one another, and are related by the equation $E = mc^2$, where c is the velocity of light.

- The *maximality of the speed of light*: the observed velocity of a physical object—that is, an object possessing mass—is always strictly less than the speed of light. This principle may be seen as an instance of the Bounded/Unbounded opposition, since before Einstein it was believed that bodies could, in principle at least, move at any speed, however large.

- A *new "relativistic" law for the addition of velocities*: if B is moving with speed u relative to A, and C is moving with speed v relative to B, then the velocity of C relative to A is $(u + v)/(1 + uv/c^2)$. Note that classically this relative velocity would be $u + v$.

Let us reflect a little on the relativistic addition law for velocities. Suppose we indicate relativistic addition of velocities by the symbol $\boldsymbol{+}$: thus $u \boldsymbol{+} v = (u + v)/(1 + uv/c^2)$. If we take $u = c$, then it is easily checked that $c \boldsymbol{+} u = c$ and by taking $u = v = c$ that $c \boldsymbol{+} c = c$. These equations are strikingly similar to the equations $\mathbf{k} + 1 = \mathbf{k}$ and $\mathbf{k} + \mathbf{k} = \mathbf{k}$ that are satisfied (as we saw in Chapter II) by an infinite cardinal number **k**. In this respect, then, *in special relativity the velocity of light, although in fact finite, behaves as if it were infinite.*

Special relativity is the source of a number of puzzles and paradoxes, each of which, in one way or another, embodies the Absolute/Relative opposition.

One puzzle that arises immediately is the following. Suppose that Jack and Jill are inertial observers in uniform relative motion. Then Jack will observe Jill's rods to be shortened and her clocks to run slow, and Jill will observe Jack's rods to be shortened and his clocks to run slow. Isn't this a logical contradiction? The answer is no. While this situation would be contradictory if Jack and Jill occupied the *same* reference frame, it is logically possible (although counterintuitive) if, as in the present case, the two occupy *different* reference frames. Jack and Jill's positions are entirely symmetric, so special relativity implies that length contraction and time dilation would be observed by *both* parties.

In the *ladder paradox* (or *barn-pole* paradox) we imagine a ladder passing at great speed through a garage with two doors, an *entrance* door and an *exit* door. When at rest the ladder is too long to fit inside the garage (with both doors closed), but the ladder's velocity is so great that, as observed by someone stationed in the garage, its length contraction enables it to fit inside the garage. On the other hand, an observer moving with the ladder would judge the garage to be moving, so that, from his point of view, it is the *garage*, not the ladder, which undergoes contraction, and is therefore unable to accommodate the ladder.

The solution to the apparent paradox lies in the relativity of simultaneity. What the garage observer judges to be two simultaneous events may not in fact be judged as simultaneous by the ladder observer. When we say the ladder "fits" inside the garage, what we mean precisely is that, at some specific time, both back and front ends of the ladder were inside the garage; in other words, both ends were inside the garage simultaneously. Since simultaneity is relative, the two observers can disagree without contradiction as to whether the ladder fits inside the garage. To the garage observer, the ladder's back and front ends were both in the garage at the same time, so that the ladder fitted in the garage; but to the ladder observer, these two events are not simultaneous, and the ladder did not fit.

This can be made clearer by allowing the garage doors to close for the brief period, *in the garage frame*, that the ladder is fully inside the garage, and then opening the doors immediately afterwards. Now consider these events *in the ladder frame*. The first event is the ladder's front end approaching the exit door of the garage. The door closes, and then opens again to allow the front end to pass through. At a later time, the back end passes through the entrance door, which closes and then opens. Since simultaneity is relative, the two doors are not required, in the ladder frame, to be shut at the same time, and so the ladder need not fit inside the garage.

There is a variant of the paradox in which the ladder is trapped once it is fully inside the garage. This can be arranged by sealing the exit door, which we assume is sufficiently robust to withstand the impact of the ladder, and then closing the entrance door at the instant when, in the garage frame, the ladder's back end enters the garage. In the garage frame, when the ladder hits the exit door, its front end instantly comes to a standstill. Since the entrance door has been closed, the ladder is trapped inside the garage. The ladder's relative velocity is now zero, so it is not length contracted, and is accordingly longer than the garage in the garage frame as well as in its own frame. As a result it will have to buckle or break.

Here the puzzle again arises when we consider the situation in the ladder frame. In its own frame, the ladder was *always* longer than the garage. So how was it possible to close the entrance door and trap the ladder inside?

Now, we have shown that, in the garage frame, the ladder is indeed trapped in the garage, and so must buckle or break. This is an *absolute* fact which must hold in *any* frame—the ladder cannot buckle or break in one frame but not in another! How can the trapping and buckling or breaking of the ladder be explained within the ladder's frame?

In the ladder's frame, the ladder is too long to fit in the garage, so by the time it collides with the exit door and stops, its back end has still not reached the entrance door. This seems to be a paradox. Does the back end cross the entrance door or not?

The difficulty here arises from the assumption that the ladder is *rigid*. Now rigidity requires that force be transmitted through the ladder instantaneously, so that, when the ladder's front end stops moving, its back end must stop moving at the same time. This would require that the force bringing the ladder to a halt be transmitted with infinite velocity. But this contradicts the maximality of the speed of light in special relativity. In special relativity, *there can be no perfectly rigid objects.*

At the instant the ladder's front end collides with the exit door, its back end is, so to speak, "unaware" of the fact that its front end has stopped, and continues to move forwards. As a result, the ladder becomes compressed. In both garage and ladder frames, the ladder's back end continues to move after the moment of collision, at least to the point where a force moving backwards at the speed of light from the point of collision could actually reach it. At this point in time, the ladder is in fact *shorter* than the original contracted length, so its back end is well inside the garage before it can even begin to stop.

What happens after the force reaches the ladder's back end? That depends. The ladder could buckle or break, or, if the garage's entrance door is not sufficiently robust, poke through it. Consistency requires only that whatever happens be the same in both garage and ladder frames.

A further variation on this paradox—in which gravity plays a role—is the *thin man and the grating*. In this scenario, a man walks so fast that the relativistic length contraction makes him very thin. He has to pass over a grating in the street. A stationary man positioned at the grating fully expects the walking man to have become so thin that he will fall under gravity through the gaps in the grating. But for the walking man it is the *grating*, and not he, who has undergone the length contraction. To him the gaps in the grating are much narrower than they are to the stationary man, and so he certainly does *not* expect to fall through them. Who is correct?

The situation can be simplified by considering a rod sliding over a flat table, in its path a circular hole of diameter the length of the rod. If the rod moves fast enough, then in the table frame its length will shrink sufficiently for it to drop easily under gravity through the hole. On the other hand, in the rod's frame, it is the diameter of the hole which shrinks, apparently making it impossible for the rod to fall through the hole.

The resolution here lies in the observation (made above) that in special relativity there are no rigid objects. In the table frame, the rod actually falls through the hole. This is an objective, frame-independent fact. Therefore the rod must also fall through the hole in the rod's frame. This is possible in special relativity because of the non-rigidity of the rod. In the moving rod's frame, it is seen to bend under gravity with its front end bending down and entering the hole and the rest of it following after! Because of the presence of gravity, the rod can look bent in one frame and not in another.

The *Ehrenfest paradox* concerns the rotation of an ideally rigid disc spinning about its axis of symmetry. Suppose that an observer *rotating with the disc* attempts to determine the ratio of the disc's circumference to its diameter, expecting the answer to be exactly π, in accordance with Euclidean geometry. He chooses a bunch of (very short) measuring rods of the same length and lays them along the diameter, and finds that n of these suffice to cover it. He then measures the circumference by laying these rods tangentially along the disc's edge, and finds that m of these suffice to cover it. On comparing m with n, he is surprised to discover that the ratio of m to n, instead of coinciding with π, actually *exceeds* it. This "paradox" can be explained as follows. First, note that if it requires a certain number of measuring rods to cover the diameter when the disc is stationary, then the *same number* of rods will suffice to cover it in the frame of an observer rotating with the disc. This is because the disc's diameter as seen in a non-rotating frame is always perpendicular to its motion and neither it, nor measuring rods laid along it, will contract. On the other hand, the same measuring rods laid along the rotating edge of the disc will undergo contraction relative to the stationary rods used to measure the diameter; *more* of the (thus contracted) rods will be needed to measure the disc's circumference in the rotating frame. It follows that, in the rotating frame, the ratio of the circumference to the diameter of the disc will exceed the ratio, namely π which would be obtained when the disc is not rotating.

This means that, for observers rotating with the disc, geometry is *non-Euclidean*. In fact, this geometry can be shown to be *hyperbolic*. This observation was important for Einstein's later development of general relativity.

In the *twin paradox* (or *clock paradox*) one of a pair of identical twins makes a journey into space in a high-speed rocket and returns to Earth to find that her stay-at-home twin is older than she is. This result appears puzzling because of the apparent symmetry of the twins' positions: the "staycationing" twin naturally regards her "vacationing" sister as travelling, but equally, the vacationing twin can view herself as staycationing (at rest in the rocket) and her sister as vacationing. An (incorrect) application of relativistic time dilation yields the paradoxical conclusion that each twin would find the other to be older.

A number of explanations of this paradox have been proposed. Most of these rest on the claim that no contradiction arises because the twins' situations are in fact not symmetric. One source of asymmetry is that, in reversing her trajectory and returning to Earth, the vacationing twin has been subjected to acceleration and deceleration, while the staycationing twin has not. This asymmetry, which involves accelerated motion, cannot receive a complete analysis within special relativity since special relativity applies only to uniform motion. It requires the tools of general relativity, which is introduced below. An asymmetry also arises from the fact that the vacationing twin must occupy *two* separate inertial frames, one on the outward journey and shifting to the other on the return, while this is not the case for the staycationing twin.

A “three-frame” or “three-clock” version of the story makes the “turn around” shift between inertial frames do the work, so avoiding issues about acceleration and keeping the analysis within the confines of special relativity. Let clock 1 “stay home” on earth while clock 2 moves. As clock 2 passes by clock 1, the two clocks are synchronized. Some distance out from home, clock 2 meets clock 3, travelling in the opposite direction. As clock 3 passes clock 2, the two clocks are synchronized. When clock 3 later passes earth, we find it reads an (aggregate) elapsed time for the return trip shorter than clock 1 does. The shift from the Einstein simultaneity relation for clock 2 to that for clock 3 “skips” a segment of time measured by clock 1.

In realistic terms, suppose the vacationing twin makes a round trip to the nearest star system beyond our solar system—a distance of 4 light years—at 80% of the speed of light. When she returns to Earth, her twin will have aged 10 years, but she will have aged only 6 years! Thus, on her return to earth, the vacationing twin has only “taken” 6 years to “travel” 10 years. This shows that, in special relativity, *future time travel is possible.*

SPACETIME

Not long after its formulation, special relativity was provided with a beautiful geometrical foundation—*four-dimensional spacetime*—by the German mathematician Hermann Minkowski. The *points* in Minkowski’s spacetime are exactly those of four-dimensional Euclidean space, and so each is specified by giving four coordinates (x, y, z, t). But while the coordinates x, y, and z are just the usual ones specifying a location in (three-dimensional) space, the fourth coordinate t is to be understood as indicating a *time.*[1] Accordingly, points in Minkowski’s spacetime are correlated, not with “locations” in an abstract *four*-dimensional space, but with *events* occurring at specific *times* and *places* in the familiar *three*-dimensional space of experience. Moreover, the distance between two events—the *interval* between them—is not calculated by means of the usual Euclidean distance formula, but in accordance with a different formula, one which takes into account the key principles of special relativity that the speed of light is independent of the state of motion of the observer and is an

1 Thus time constitutes the “fourth dimension” in Minkowski spacetime.

upper limit to all physical velocities. In fact, the interval between two events ***e*** and ***e***' with coordinates (x, y, z, t) and (x', y', z', t') is given by

$$(1)\quad d(\boldsymbol{e}, \boldsymbol{e}') = \sqrt{c^2(t'-t)^2 - [(x'-x)^2 + (y'-y)^2 + (z'-z)^2]}$$

where c is the velocity of light.

To understand the meaning of this equation, imagine that a flash of light emanates from a point P with spatial coordinates (x, y, z). If $x' = x + \Delta x$, $y' = y + \Delta y$, $z' = z + \Delta z$, $t' = t + \Delta t$, then (1) may be put in the form

$$(2)\quad d(\boldsymbol{e}, \boldsymbol{e}')^2 = c^2(\Delta t)^2 - [(\Delta x)^2 + (\Delta y)^2 + (\Delta z)^2].$$

Now $(\Delta x)^2 + (\Delta y)^2 + (\Delta z)^2$ is the squared spatial distance between P and the point P' with coordinates (x', y', z'): this quantity is called the *spatial part* of the (squared) interval between ***e*** and ***e***'. The term $c^2(\Delta t)^2$ is the squared distance from P of the wavefront of our flash of light at time $t + \Delta t$: this is called the *temporal part* of the (squared) interval between ***e*** and ***e***'. Accordingly the interval between two events is the square root of the *difference* between its temporal and spatial parts. If this difference is *positive*, the interval between or *separation of* the events is called *timelike;* if zero, *lightlike*; and if negative, *spacelike.* A timelike separation of the events ***e*** and ***e***' indicates that a light ray emanating from the location at the time of occurrence of ***e*** would traverse—barring obstacles—the spatial distance to the location of ***e***' before ***e***' actually occurs. So the event ***e*** could, in principle, give rise to a physical influence (such as that produced by the motion of a particle, which according to special relativity must necessarily travel at less than the speed of light) which moves sufficiently quickly so as to "cause" or "trigger" the event ***e***'. Similarly, a lightlike separation means that a light ray originating at the time and spatial location of ***e*** could trigger the event ***e***'. A spacelike separation, on the other hand, would mean that the event ***e*** could only trigger the event ***e***' through an influence moving faster than light, which, according to the theory of relativity, is impossible. This is reflected in the fact that the interval associated with a spacelike separation would, be as the square root of a negative number, an imaginary quantity.

It is illuminating to map out in a "spacetime diagram" the location of events that can be connected by a light ray to a given event ***e***. For simplicity let us suppose that ***e*** takes place at the origin (0, 0, 0, 0) of the spacetime diagram. Then if the spatial coordinates of the event ***f*** are (x, y, z), its time coordinate t either has the value

$$(3)\quad t_{\text{future}} = +\sqrt{x^2 + y^2 + z^2}$$

or

$$(4)\quad t_{\text{past}} = -\sqrt{x^2 + y^2 + z^2}$$

The graphical presentation of this formula is simplified by confining attention to events ***f*** whose z-coordinate is zero. In that case the spacetime diagram may be presented as if it had just two spatial coordinates x and y together with the time coordinate t:

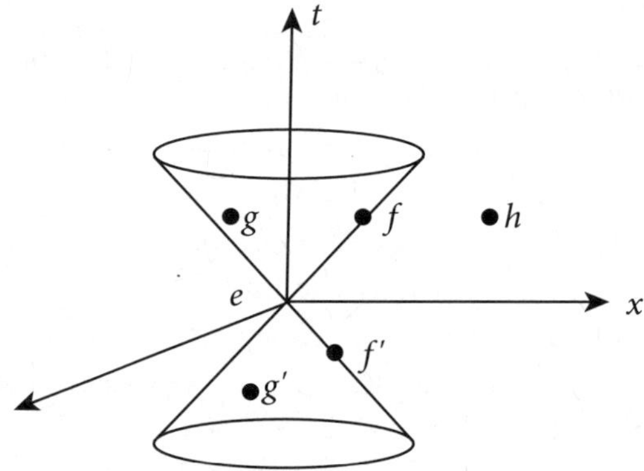

Each event in this diagram with lightlike separation from ***e*** either lies, like ***f***, on the surface of the *future light cone* of ***e*** (equation 3) or, like ***f***', on the surface of the *past light cone* of ***e*** (equation 4). An event such as ***g*** contained within the future light cone can be caused by ***e***, and one such as ***g***' in the past light cone can be the cause of ***e***. On the other hand, an event such as ***h*** which is entirely outside ***e***'s light cone cannot be causally related in any way to ***e***: it is *causally independent* of ***e***.

In this way the special theory of relativity made possible a remarkable fusion of the opposed concepts of space and time.

FASTER-THAN-LIGHT PARTICLES IN SPECIAL RELATIVITY: TACHYONS

It is a consequence of special relativity that the velocity of any *physical* object, that is, any object possessing mass, must always be strictly less than the speed of light. Essentially, this is because any attempt to accelerate a physical object to the speed of light would cause its mass to become infinite. But *massless* particles such as photons and neutrinos do move at the speed of light. Surprisingly, it is consistent with special relativity that "particles" exist which travel at *superluminal* velocities, that is, at velocities *exceeding that of light*. Such "particles" are called *tachyons*, from the Greek *tachis* ("swift").

Tachyons, if they existed, would have very strange properties. For example, the mass of a tachyon would have to be represented by an imaginary number, that is, a multiple of $\sqrt{-1}$. Moreover, as tachyons lose energy they would *accelerate* rather than decelerate as would an ordinary particle. Similarly, if a tachyon receives energy, it *decelerates.*

Tachyons would also possess the disturbing property of appearing to *travel backwards in time*. We recall that the time dilatation effect causes moving clocks to slow down, and as the speed of the clock approaches that of light, the observed rate at which the clock runs run tends to zero. That being the case, a tachyonic "clock" travelling at a superluminal velocity relative to an observer would appear to be running backwards. This would appear to open the possibility of using tachyons to communicate with the past. But such a possibility results in paradox. For suppose that Jack and Jill enter into the following agreement: Jack will send a message to reach Jill at four o'clock provided he does *not* receive one from Jill at two o'clock. Jill agrees to send a message to reach Jack at two o'clock immediately on receiving one from Jack at four o'clock. In that case, the exchange of messages will take place if and only if it does not take place. This is a contradiction.

We conclude that Jack and Jill cannot, on pain of contradiction, enter into such an agreement. Does this mean that the existence of tachyons is ruled out on logical grounds? No, provided we accept, as in the case of time travel into the past, that any attempt at tachyonic communication with the past causes the timeline to "split" into two at the time of reception of the message. The latter is then received, not by the intended recipient in the original timeline, but by his or her counterpart in the new timeline. In the case at hand, Jack's message would be sent into the future and so could reach the original Jill. On the other hand Jill's message would be sent into the past and so necessarily could not reach the original Jack, but only his counterpart Jack 2 in the other branch of the split timeline. Thus Jack will *fail* to receive a message from Jill, so causing him to send her a message. This in turn will cause Jill to send a message, but it will be received by

Jack 2, and *not by Jack himself*. So Jack will not be forced to generate a contradiction by failing to send a message to Jill.[2]

Just as in the case of time travel into the past, the question of the existence of tachyons remains open.

GENERAL RELATIVITY: THE PRINCIPLE OF EQUIVALENCE

The special theory of relativity concerns only uniform, constant, unaccelerated motion. In his *general theory of relativity*, Einstein extended the special theory to accelerated, changing motion, more particularly, to motion under gravity. The move from special to general relativity thus embodies the opposition between the Constant and the Changing.

The key idea on which general relativity rests is the *Principle of Equivalence*.

In his celebrated book *Relativity: The Special and the General Theory*, first published in 1916 and still in print, Einstein uses a parable to explain the Principle of Equivalence. He imagines "a spacious chest resembling a room with an observer inside who is equipped with apparatus" deep in space. The chest is situated far from any gravitational fields, so far that any object moving uniformly in a straight line will continue on indefinitely. Since no gravity is present, the observer will drift around the room, weightless. Now Einstein imagines a rope attached to a hook in the centre of the chest's top and introduces an unspecified "being" who proceeds to pull the rope with a constant force. As a result, the chest and its contents, including the observer, accelerate "upwards" at a constant rate. What does the man in the chest make of this? No longer adrift, he finds himself moving towards what is now to him the "floor" of the room. Once he lands there, he needs to stiffen his legs to stand. Anything he drops now moves "downwards" in acceleration towards the floor. He also finds that all bodies accelerate at the same rate. He infers from all this that the chest is in a gravitational field. Then he asks himself: why doesn't the room itself fall? At this point he discovers the rope attached to the hook, and concludes that it is the rope which prevents the room from falling.

2 We have touched on the strange consequences of tachyonic communication in our discussion of Gregory Benford's *Timescape* in the previous chapter.

Now an observer outside the chest would disagree. She would deny that a gravitational field is present, and would claim that the experiences of the observer in the chest are the result of the mysterious "being" pulling the chest upwards. Who is right? The Principle of Equivalence asserts that, in a certain sense, *both are right*: the point of view of the observer in the chest is no less valid than that of the external observer. To be precise, *being inside the uniformly accelerating chest is physically equivalent to being in a uniform gravitational field.*

Now suppose that, instead of being pulled "upward" the chest is freely falling in a gravitational field. In that case, everything inside it will be weightless. Thus the observer in the chest will conclude that the chest is deep in space, far from any gravitational fields, while an external observer will see that the chest is freely falling in a gravitational field. Again, the Principle of Equivalence asserts that, in a certain sense, both are right: what the observer in the chest reports is no less valid than the report of the external observer. To be precise, *being inside the freely falling chest is physically equivalent to being far away from any gravitational field and undergoing no accelerations.*

To sum up, the Principle of Equivalence asserts that:

- The physical experience of an observer undergoing uniformly accelerated motion is indistinguishable from that of an observer stationary in a gravitational field.

- The physical experience of an observer falling freely in a gravitational field is indistinguishable from that of an observer far from any significant source of gravity. Another way of putting this is to say that *any reference frame freely falling in a gravitational field is inertial.*

The equivalence between a gravitational field and acceleration is possible only because gravitational mass, or weight, is exactly equal to inertial mass, or resistance to acceleration. Since a gravitational field affects all masses in the same way, it can be "cancelled out" by freely falling with it. Masses then become weightless just as do the astronauts in a space station orbiting the earth. This is a property unique to gravitational fields. There is no way of cancelling out an electric or magnetic field by "freely falling" with it, since mass does not determine electric or magnetic charge uniquely. Remarkably, mass *does* determine "gravitational charge"—or weight—uniquely.

The Principle of Equivalence has several immediate, and startling consequences.

To begin with, suppose that a beam of light is aimed through a hole in the wall of the uniformly accelerating chest. Because of the acceleration, the observer in the chest will see the path of the beam of light as slightly curved, or "bent." He will regard this bending as an effect of the gravitational field he believes is present. By the principle of equivalence, his conclusion may be considered as correct. It follows that *light bends in a gravitational field*. This effect was first detected in 1919 by the English astrophysicist A.S. Eddington, who observed stars close to the sun during a solar eclipse. Newspaper reports of the event led to Einstein's enduring fame.

Now suppose that the beam of light is aimed upwards through a hole in the chest's floor. In the interval between the time the beam is shot from the floor and the time it reaches the ceiling the chest's acceleration will have caused the chest's velocity to increase. Thus the observer (supposed suspended from the ceiling) will see that the light is redshifted to a lower frequency owing to the Doppler effect.[3] Again, he will attribute this redshift to the presence of a gravitational field. The principle of equivalence then implies that *light* (or any electromagnetic radiation) *is redshifted as it moves upwards in a gravitational field*.

An even more remarkable consequence is this. Suppose that the light is emitted as the result of oscillations in an atomic clock. Then the redshifting of the light can be interpreted as a *slowing down of the clock*. Consequently, *in a gravitational field, time passes at different rates in different locations: the stronger the intensity of the field, the slower time passes*. The US atomic standard clock, housed at 5400 feet in Boulder, Colorado, gains 5 microseconds per year over an identical clock situated almost at sea level in the Royal Observatory at Greenwich, England. (Both clocks are accurate to one microsecond per year.) So someone living on the surface of a planet with a strong gravitational field would age more slowly than someone on the earth's surface.

---

3 The *Doppler effect* is the change in frequency of a wave (or other periodic event) for an observer moving relative to its source. If the observer is moving towards the source, the frequency will appear to increase, while if the observer is moving away from the source, the frequency will appear to decrease. In the case of a source of light, a motion away from the source will cause the apparent colour to tend toward the red end of the visible spectrum; this is known as a *redshift*.

We have seen that, by the Principle of Equivalence, light bends or curves in a gravitational field. In classical Newtonian physics, this phenomenon is explained by asserting that gravity exerts a force of attraction on light just as it does on every physical object. Einstein's explanation was quite different, and indeed revolutionary. He postulated that gravity is not a force of attraction at all, but is, rather, a *curvature in the fabric of spacetime*. In a gravitational field, spacetime itself is "warped" or "bent" in such a way that the equivalent of straight line paths—the paths of shortest length or *geodesics*—are not Euclidean straight lines but *curves*. The paths of light rays are always geodesics, but in the curved spacetime of a gravitational field these geodesics are curves, just as geodesics on the curved surface of a sphere are great circles rather than Euclidean straight lines. Newton's first law of motion asserts that that an object will remain at rest or in uniform motion in a straight line unless compelled to change its state by the action of an external force. In Einstein's gravitationally curved spacetime this law continues to hold with the term "straight line" replaced by "geodesic." An object falling freely in a gravitational field experiences no "force" on it and so therefore must be moving in a geodesic path. A satellite in orbit around the earth is freely falling in the earth's gravitational field and so its trajectory must be a geodesic, i.e., a "straight line" in the corresponding curved spacetime. From the point of view of an observer on the earth the satellite's orbit is curved, but from the point of view of the satellite, its orbit is a path of shortest length, i.e., "straight." This is a striking instance of the opposition between the Straight and the Curved.

The spacetime of general relativity is a Riemannian manifold with variable curvature. It can be compared to a sheet of rubber which can be stretched or twisted. The presence of a material body has the effect of stretching and twisting the spacetime in its vicinity, causing nearby objects to travel in curved paths. This distorted spacetime in an object's vicinity *is* the object's gravitational field. The physicist John Wheeler has expressed this memorably: *matter tells spacetime how to curve, and curved spacetime tells matter how to move.*

BLACK HOLES

One of the most mind-bending concepts in general relativity—a concept which has captured the popular imagination—is that of a *black hole*. A black hole is a region of spacetime from which nothing, not even light, can escape.

The idea of a black hole is closely connected with the idea of *escape velocity*. If a stone is thrown upwards, then the faster it is thrown, the higher it rises and the farther it travels before falling back. If it is thrown sufficiently fast, it will overcome the earth's gravitational pull and escape, never to return. The velocity an object must have to overcome a mass's gravitational attraction in this way is called the mass's *escape velocity*. In the case of the earth the escape velocity is about seven miles per second.

Now imagine the earth being steadily compressed into a smaller and smaller volume. This causes its density to increase. As a result, the gravitational pull at its surface—and so also its escape velocity—increases. Eventually there comes a point when the degree of compression, and the corresponding gravitational pull, become so great that the compressed earth's escape velocity exceeds that of light. At this point, nothing can escape, since no physical object can travel faster than light. The result is a black hole.

Just as with ordinary astronomical objects, the gravitational attraction of a black hole at very large distances is negligible. But as it is approached, its gravitational pull, and associated escape velocity, steadily increases until, at some point, that escape velocity reaches the speed of light. The region surrounding a black hole at which this occurs is called its *event horizon*. Light emitted from inside the event horizon can never reach an observer outside it.

It may well be asked: when a body is compressed into a black hole, while the black hole retains the gravitational attraction of the body, what happens to the body itself? General relativity predicts that it will be compressed into an infinitely dense mass with zero volume, known as a *singularity*, situated at the black hole's centre. At a singularity, the laws of physics as currently understood break down.

Now imagine an astronaut travelling in a spacecraft towards a black hole's event horizon, watched by an observer positioned far from the black hole. What will the observer see? As the astronaut approaches the black hole's event horizon, and the local gravitational field increases in intensity, the slowing of time in gravitational fields will make it appear to the external observer that the astronaut is moving more and more slowly. This retardation of time prevents the external observer from actually seeing the astronaut reach, and pass through the event horizon.

To the external observer, the astronaut's image gradually "freezes" at the event horizon; she would have to wait forever to see the astronaut actually reach it. From the observer's point of view, the passage of time in the astronaut's spacecraft slows to a virtual standstill.

Yet the astronaut himself experiences no such untoward temporal effects. In his frame of reference, he arrives and passes through the event horizon in what for him is a finite amount of time. Now if he remains in communication with the observer before reaching the event horizon, he will see the latter's clocks speeding up, approaching infinity as he gets closer to the event horizon. He will consider that, paradoxically, at the instant he passes over the event horizon, an *infinite* amount of time has passed in the external observer's reference frame.

What does our astronaut experience after passing through the black hole's event horizon? That depends on the type of black hole. In the case of a nonrotating black hole, the astronaut will be drawn inexorably to the singularity at its centre—the infinitely dense residue of the mass whose compression originally produced the black hole. As he approaches the singularity his body is subjected to increasing tidal forces arising from differences in the strength of gravity. The closer a point is to the singularity, the stronger the gravitational pull there. Thus, if the astronaut were to fall head-first into the black hole, then, as he approaches the singularity, there will be an increasing difference in the gravitational acceleration between his head and his feet which will eventually wrench his body apart. These tidal forces can become very large even outside the black hole's event horizon. For example, a small black hole with the mass of the sun has a radius of 3 kilometres, yet the tidal force it would exert across our astronaut 100 kilometres away is a truly staggering 51700 earth gravities. The poor man would be stretched to the thinness of capellini. On the other hand, a large black hole with mass 100 million times that of the sun has a radius of 300 million kilometres, but the tidal force it exerts on an astronaut 100 kilometres distant is negligible.

Once our astronaut has traversed a large black hole's event horizon, how long does it take for him—in free fall—to reach the singularity? In the case of the 100 million solar mass black hole, it would take about 25 minutes. For a truly monstrous 21 billion solar mass black hole, the corresponding time is somewhat less than 4 days.

As our astronaut passes over the event horizon of a monster black hole, say the one above of 21 billion solar masses, he notices nothing special. It is likely that he is blissfully unaware that he has entered a black hole, and so has no idea that he has less than 4 days to live. And, surprisingly perhaps, as he continues his ill-fated journey to the singularity, he won't actually see anything particularly novel, at least in his immediate vicinity. In particular, he will continue to be able to see objects outside the black hole, since light from outside the black hole can still reach the inside. One visual oddity he might notice is the curious distortion of the appearance of remote objects owing to the bending of light by the black hole's gravity. In fact, our astronaut might not notice anything until he becomes uncomfortably aware of the tidal effects on his body when he gets sufficiently close to the singularity. Soon afterwards he is torn apart.

The fate of an astronaut having the misfortune to fall into a (nonrotating) black hole is thus completely sealed. There is, according to general relativity, no way whatsoever, even in principle, of avoiding his eventual encounter with the singularity. He may be able to delay his fate by firing the jets of his spacecraft to counteract his fall, but no matter how much energy is expended, after a finite amount of time he will be sucked into the singularity. The route to the singularity *must* be travelled, just as we are pulled ineluctably into the future. Chapter IV began with the observation that space differed from time in that it is possible to move about freely in space, but not (apart from the development of time machines) in time. It was further remarked that this is so because space is *isotropic*—there are no privileged locations in space. But in the interior of a black hole there *is* a privileged location—the singularity. In the ancient world, all roads led to Rome. In a black hole, all roads lead to the singularity. Just as time pulls everything remorselessly into the future, so in a black hole everything is pulled equally remorselessly into the singularity. So, in a sense, a black hole transforms space into time.

These observations apply to nonrotating black holes. But *rotating* black holes offer the possibility of other, happier scenarios. An astronaut travelling into a large rotating black hole could, in principle, contrive to avoid the singularity and then pass—through a so-called "wormhole"—into another region of spacetime totally disconnected from the one from which he embarked—in a word, another "universe."

These strange, seemingly impossible phenomena can be exploited to enable—in principle at least—the completion of supertasks.

For suppose we imagine our astronaut approaching the event horizon of a large rotating black hole, viewing a clock positioned in a reference frame far from the black hole. He would see the rate at which the distant clock runs speed up indefinitely. At the instant he arrives at the black hole's event horizon, he will judge that, according to the distant clock, an infinite amount of time has passed. In that case, supertasks in the distant reference frame would be regarded by the astronaut as finite, completable tasks. Thus, supposing the astronaut to be an amateur mathematician, he could learn the truth or falsity of undecided propositions in number theory asserting that every natural number has a certain property (e.g., Goldbach's conjecture that every even number is the sum of two primes) by receiving information from a distant computer checking each natural number for the property in question. He could even determine whether mathematical theories such as axiomatic set theory are consistent (which it is known cannot be established by finite means) by having the distant computer run through all the possible proofs in the theory to check whether an inconsistency shows up. The astronaut could arrange to receive the relevant information by having the computer transmit a signal (a beam of electromagnetic radiation of a prescribed frequency, for example) to him if it turns up an even number not the sum of two primes or if it finds a proof of a contradiction in axiomatic set theory. If by the time the astronaut passes through the event horizon of the black hole he has received no signal from the computer, this means that the computer, in its infinite search, has found no exception to Goldbach's conjecture or no inconsistency in axiomatic set theory, so that he can be confident Goldbach's conjecture is true or axiomatic set theory is consistent. If on the other hand he does receive a signal, this would mean that Goldbach's conjecture is false or axiomatic set theory is inconsistent. The distant computer might have to run forever in its own reference frame but to the astronaut the time taken to arrive at a conclusion would always be finite.

If our astronaut has the good fortune to survive passage through the black hole, he will pass through a wormhole to emerge in a different "universe" from the one both he and the computer originally occupied. From the point of view of the astronaut, now occupying the "new" universe, an infinite amount of time will have passed in the "old" universe he left.

It is amusing to speculate on the experiences the astronaut might have in this new universe. Supposing that universe to contain mathematicians, the astronaut could present himself to them, explain how he arrived in their universe and go on to claim that he knows for a fact that, say, Goldbach's conjecture is true. The alien mathematicians (assuming that they have not themselves proved or refuted Goldbach's conjecture) might, of course, simply dismiss the astronaut's claims as the ravings of a lunatic. But what if they were to believe his story? In particular, suppose that they accept his claim that he received no signal from the computer. Would they then, along with the astronaut, accept this as decisive evidence for the truth of Goldbach's conjecture? Probably not, for in most mathematicians' view the only way to establish the truth of a mathematical proposition such as Goldbach's conjecture is to provide a *proof* of it. The astronaut's "evidence," consisting merely of his claim of failing to receive a signal from a computer, provides no means whatsoever of constructing a proof of Goldbach's conjecture. Moreover, even the reliability of the procedure, taken at face value, is questionable since it depends on the exotic physics of black holes.

Fascinating as all this is, it is intellectual speculation. Do black holes truly *exist* in a robust physical sense, in the same sense as we take, say, the earth or the moon to exist? Since no light can escape from a black hole, it is impossible to see one directly. But black holes do exert a gravitational influence, and that *is* detectable. Also there are good scientific reasons for supposing that they exist. According to contemporary astrophysics, when a star of sufficient size has exhausted all of its fuel, it contracts and as a result heats up sufficiently to explode into a supernova. The matter remaining after the explosion is sufficiently dense to contract ("self-compress") under its own gravitational attraction to a *neutron star*—a superdense object, so-called because its gravitational self-attraction is so powerful that all of its constituent protons and electrons have been forced together to become neutrons. Five hundred or so neutron stars have actually been identified through the use of radio telescopes.

If the resulting neutron star is sufficiently massive, general relativity predicts that its self-gravitation will overwhelm the forces preventing its constituent neutrons from being mashed together. As a result the neutron star will shrink inexorably until it ends up as a black hole. For this to happen, the mass of a neutron star need only be about twice the mass of the sun, so that, the sun not being a particularly large star, it is reasonable to expect at least a few neutron stars to be this massive.

Astrophysicists believe that, on average, a supernova should be produced at least every 300 years. Given the immense age of the universe, it is likely that some of these supernovas have produced black holes. Recent observational evidence also supports what astrophysicists have suspected for some time, namely, that the centres of galaxies—including the Milky Way—contain supermassive black holes formed by the presence of millions, even billions of stars packed into the comparatively small space at the galaxy's core. That being the case, the crew of any future actualization of Star Trek's *USS Enterprise* should beware! They may pass unwittingly over the event horizon of a supermassive black hole, and that, sadly, will be the end of them.

Chapter VI

Puzzles and Paradoxes in Quantum Physics

Quantum physics, quantum mechanics, or quantum theory, is one of the cornerstones of modern physics. Yet from its beginnings in the early twentieth century it has bristled with puzzles and paradoxes. While these difficulties originated in the opposition between the Continuous and the Discrete, quantum theory also offers striking instances of the opposition between Chance and Necessity, the Absolute and the Relative, and the Whole and the Part.

WAVES VS. PARTICLES

A moving body does not "jump" discontinuously from one place to another but (we suppose) passes over all places in between. In the same way, when a body falls under gravity, its velocity and kinetic energy increase continuously, and when a body is heated, its temperature rises continuously. Phenomena such as these suggest that *physical change is always continuous*. This, the *Physical Continuity Principle*, is a basic tenet of classical physics. What it really amounts to is that the measured value of any quantity associated with physical change also varies continuously.

It is true that some physical quantities are *discrete*, for example the number of faces in a crystal, and the atomic number of an atom. But, unlike speed, temperature, or energy, these are fixed structural features of the things in question and cannot be changed without inducing a fundamental change in the things themselves. Changing the atomic number of an atom from 6 to 8 would transform an atom of carbon into an atom of oxygen.

The Physical Continuity Principle stood unchallenged until 1900 when the German physicist Max Planck put forward what came to be termed the *Planck postulate*, one of the fundamental principles of quantum physics. Planck was led to formulate this principle in the course of efforts to resolve the problem of *black-body radiation*. This is the question of how the intensity of the electromagnetic radiation emitted by a perfect absorber of radiation—a so-called black-body—depends on the frequency of the radiation (in the case of light, its colour) and the temperature of the body. The problem had previously been investigated experimentally, but no theoretical treatment agreed with the experimental values.

After a number of unsuccessful attempts at solving the problem, in 1900 Planck formulated his *black-body radiation law* which was closely in accord with what had been experimentally observed. In deriving this law he made the crucial assumption—the *Planck or quantum postulate*—that the energy associated with electromagnetic radiation of a given frequency could be emitted by a black body only in discrete units, or *quanta*, whose value depends in a simple way on the frequency of the radiation. Planck postulated the existence of a constant h—which came to be known as *Planck's constant*—such that the energy of a quantum associated with black-body radiation of frequency ν is $h\nu$. If there are n quanta, the total energy E of radiation of frequency n is then given by $E = nh\nu$. The energy of radiation is thus directly proportional to the frequency and inversely proportional to the *wavelength*, defined as the reciprocal of the frequency.

At the time he formulated his quantum postulate, Planck was, along with his contemporaries, still wedded to classical physics. Uncomfortably aware that his postulate was incompatible with the Physical Continuity Postulate, he regarded the introduction of quanta as a purely formal procedure, a "mathematical trick." He did not foresee that this "trick" would come to be celebrated as the birth of a new physics—*quantum physics*—in which the classical assumption of continuity is replaced by discreteness, and which would lead to a profound change in our understanding of the physical world.

Planck viewed his quanta, or *photons*, as nothing more than mathematical fictions. But in 1905, Einstein, in another of his epoch-making papers, made the bold suggestion that photons, far from being fictions, were *real physical objects* with *discrete*, as opposed to continuous energies. Einstein's paper concerned the *photoelectric effect*. This is the phenomenon that certain metals can be caused to emit electrons when exposed to a beam of light (this effect was to provide the basis for the photoelectric cell). The effect was presumed to result from the transfer of energy from the light to electrons in the metal. According to classical theory, light and other forms of electromagnetic radiation consist of continuous waves, so that an alteration in either the intensity or the frequency of the light should induce a change in the rate of emission of electrons from the metal. Moreover, on this reckoning, a sufficiently dim light would be expected to show a lag time between the initial shining and the subsequent emission of an electron. But the observed behaviour of the emitted electrons is very different from what the classical theory predicts. For example, electrons are only dislodged by the photoelectric effect if the frequency of the light attains a certain value, below which no electrons are emitted regardless of the intensity of, and time of exposure to, the light. To make sense of this, Einstein proposed that the beam of light impinging on the metal is not a wave propagating through space, but rather a collection of *discrete photons*, each with energy $h\nu$. In order to dislodge an electron in the metal, a photon must have a certain minimum energy. No matter how many photons with energies below that minimum impinge on the metal, no electrons will be dislodged.

The claim that light consists of particles, or *corpuscles*, had been put forward by Newton in 1704 in his work *Opticks*. Newton's corpuscular theory remained largely accepted for nearly a century after it was proposed. But in 1803 Thomas Young's demonstration of optical interference showed decisively that light must consist of *waves*. He achieved this by performing the celebrated *double slit experiment*.

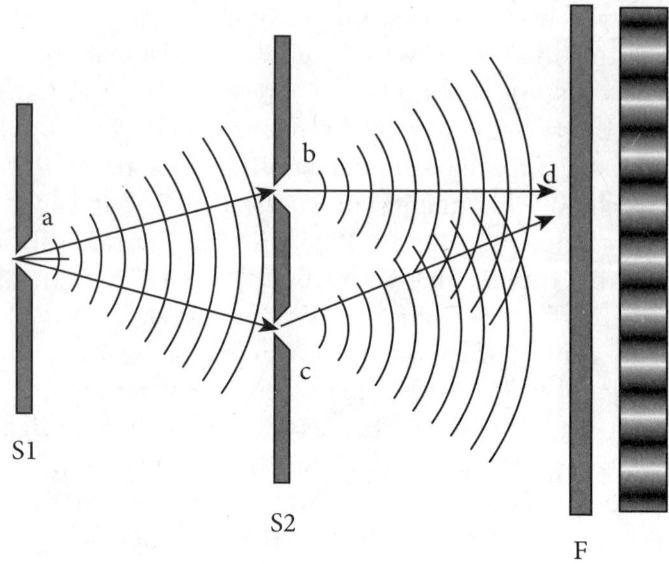

In Young's experiment, a beam of light is made to pass through a narrow aperture and then through a pair of closely spaced slits to form an image on a screen. By passing through the pair of slits, the beam is split into two. The image formed on the screen is seen to form an interference pattern, consisting of evenly spaced bright and dark bands. This result would not be expected if light consisted of classical particles, but is easily explained if light is composed of waves. Young correctly traced the origin of the interference to the differences in the path lengths of the two sets of waves. Some waves would arrive in phase to create the bright bands, while others would arrive out of phase, leaving dark bands on the screen.

The double slit experiment seemed to have provided decisive evidence that light (and electromagnetic radiation generally) is wave-like and so continuous in nature. Yet the action of light in the photoelectric effect could only be explained by supposing that a beam of light consists of individual photons and so is discrete. This seems little short of a contradiction. How can something be *both* continuous and discrete?

Now the contradiction might be resolved by noting that the double slit experiment, in which light behaves like a continuous wave, and the photoelectric effect, in which light behaves like a collection of discrete particles, are conducted under entirely different physical arrangements. But in fact the wave-like and corpuscular behaviour of light can be exhibited in a *single* experimental setup.

This can be done by refining the double slit experiment. We suppose that the slits can be opened and closed, and that the light source is so weak that *no more than a single photon passes through each slit at a time.* This eliminates the possibility of interaction between them. We also suppose that the screen is a photographic plate so that each photon striking it leaves a definite, dark imprint or "spot." As it strikes the screen, the beam of light thus *behaves in a purely corpuscular way.*

Now consider what takes place in the space between the slits and the screen. Three cases arise, according as just one of the slits is open, or both are. (When both slits are closed, no photons get through to the screen.)

First, suppose that slit 1 is open and slit 2 is closed. In that case the pattern of photon impacts on the screen turns out to be the same as the diffraction pattern which would be produced by a wave passing through the single slit, that is, a roughly circular patch centred directly behind the open slit, dark at the centre and gradually lightening as the distance from the centre increases. (Recall that the screen is a photographic plate which is darkened by light impinging on it.) In the second case, with slit 1 closed and slit 2 open, the same pattern is observed, except that the dark patch is now centred directly behind slit 2.

What is observed when *both* slits are open? If the photons were "ordinary" particles, like grapeshot, say, one would expect the pattern produced on the screen to be the sum of the two single slit patterns. This is very far from what is actually observed. There is a point of maximum darkness, located at the point on the screen directly behind the centre of the gaps between the two slits. And, unlike the sum of the two single-slit patterns, numerous other points of maximum darkness can be observed as well. Even more remarkably, in between these points—the *local maxima*—regions nearly free of photon impact can be seen, in positions where the sum of the two single-slit patterns is not zero! That is, there are locations where the two open slits produce *fewer* impacts than each slit separately. At these locations *destructive interference* of the photons has taken place. On the other hand, local maxima are locations where the number of photon impacts is *greater* than the sum of the two single-slit patterns. Here *constructive interference* has occurred. The pattern produced by the two open slits, is, in fact, precisely the interference pattern observed when light is considered to be a continuous wave.

The interference of the photons cannot arise as the result of an interaction between them, since the experiment is set up in such a way as to eliminate the possibility of any such interaction. But could such wave-like behaviour be the result of individual photons somehow splitting into two and passing through *both* slits? Experiment shows that this is not the case. If detectors are placed immediately behind both slits, then one can tell which slit each photon passes through, or whether it splits and passes through both. No evidence is ever found for photon splitting.

The fact that the photons leave a point-like trace as they impinge on the screen can be explained by assuming that they are discrete particles. That they exhibit diffraction in the single-slit case and self-interference in the two-slit case become explicable if they are pictured as continuous waves. *But the results of the whole experiment cannot be explained by consistently assuming them to be one or the other.*

Here is another way of looking at the problem.

If the photons were discrete particles, whether the slit the photon fails to pass through is open or closed should not matter, and accordingly the two-slit pattern should simply be the sum of the two single-slit patterns.

On the other hand, if the photons were "really" continuous waves, or bundles of waves, one is left with the problem of explaining their pointlike, discrete behaviour as they impinge on the screen. It would seem that there is no alternative but to conclude that in some mysterious way they 'collapse' into particles as they hit the screen.

Remarkably, if one attempts to determine *which* of the two slits each photon passes through (for example, by placing a "detector" next to each slit), the effect is to transform the double-slit interference pattern into the sum of the single-slit patterns. The more precise the determination of which slit each passes through, the closer the pattern on the screen comes to resemble the sum of the single-slit patterns. Pushing this to the limit, it is possible to install a shutter which oscillates up and down in such a way that only one shutter is open at a time. In that case it is known precisely which shutter (the open one) each photon passes through, and the pattern on the screen is exactly the sum of the two single-slit patterns.

The fact that processes such as two-slit interference demand that light display both wave-like, continuous, and corpuscular, discrete properties has come to be known as *wave-particle duality*. This is a strikingly paradoxical instance of the opposition between the Continuous and the Discrete.

Wave-particle duality applies not just to light, but also to quantum objects such as electrons which were initially identified as particles. Experiments have shown that electrons can exhibit the same kind of interference patterns as does light.

Discreteness is a characteristic feature of the quantum world. In classical mechanics planets or asteroids orbiting a star can have (in principle at least), arbitrary rotational velocities, and so a continuous range of kinetic energies. But according to quantum mechanics, electrons bound to an atom can only occupy a discrete set of states associated with a certain discrete distribution of energy levels. When an electron in an atom changes its energy level by passing to a different state, it does so *discontinuously* by means of a so-called quantum leap. If the change in energy is negative, the electron emits a photon of a certain energy, so losing exactly that amount of energy in the process. The process is *indivisible*; the electron's change of state can never yield fractions of a photon.

HEISENBERG'S UNCERTAINTY PRINCIPLE AND BOHR'S PRINCIPLE OF COMPLEMENTARITY

An important consequence of wave-particle duality is that it imposes limits on the quantity of information about a quantum system that can be obtained at any one time. In the double-slit experiment, for example, we have the option of measuring the wave properties of light by allowing it to pass through the double slit without determining which slit each photon passes through, leading to the interference pattern; or of observing the photons as they traverse the slits. *But both cannot be done at the same time.*

Werner Heisenberg, a physicist who played a central role in the early development of quantum physics, realized that the impossibility in principle of performing a simultaneous measurement of the wave properties and particle properties of light could be interpreted in a different way. He pointed out that the determination of which slit a photon passes through is essentially a measurement of the *position* of the photon as it passes through the double slit. The observation of the interference pattern can be viewed as essentially a measurement of the *direction* of the photon's trajectory, and hence of its *momentum* as it emerges from the slits. Wave-particle duality therefore leads to the conclusion that it is impossible to make simultaneous and completely precise position/momentum measurements on a quantum object such as a photon or an electron. This is *Heisenberg's uncertainty principle*. In mathematical terms, he showed that the product of uncertainty in the position of a particle by the uncertainty in its momentum *can never be smaller than the Planck constant*.

Later a "time/energy" version of the uncertainty principle was formulated. This asserts that the more precisely the time at which a particle's energy is measured, the less definite is the result of that energy measurement. In particular, if one attempts to measure the energy of a particle at a precise time, the result will be completely indeterminate.

Heisenberg provided a nice intuitive explanation of position/momentum uncertainty. To determine the position of a particle such as an electron, the simplest method is to shine a beam of light on it. Treating light as a wave, the light waves in the beam illuminating the particle will be scattered in all directions. Some of these waves will be detected by the observer's eyes, or by the measuring apparatus, thus producing an image of the particle which enables the observer to determine—to some degree of accuracy—the particle's position. Because of the wave nature of light, this image cannot be perfectly sharp. Features smaller than the wavelength of the illuminating light are necessarily blurred and indefinite, and it follows that the particle's position cannot be measured more accurately than the light's wavelength. Now to cope with this difficulty, one could illuminate the particle with light of shorter and shorter wavelengths. But here the corpuscular nature of light enters the picture. As we have seen, light consists of photons whose energy is inversely proportional to the wavelength. To illuminate the particle with light of a very short wavelength is thus to bombard it with photons of a very high energy. The impact of these photons causes the particle to recoil, so changing its velocity in a sudden and unpredictable way. This is the source of the uncertainty: the greater the desired accuracy in a position measurement of a particle, the shorter the wavelength of the light that should be used, and, in turn, the greater the resulting impact on the particle's velocity. To put it succinctly, *the very act of observation affects the observed*.

The major consequence of the uncertainty principle is that a particle can no longer be said to possess a distinct, well-defined position and velocity, but only a *quantum state*, a kind of combination of position and velocity. When the precise values of all of the properties of a physical system such as an electron cannot be known at the same time, those properties which are not known with precision must be assigned, not definite values, but *probable* values, or *probabilities*. According to quantum theory, these probabilities are all that can be known in advance about physical systems subject to the uncertainty principle. This conclusion demolished the strict determinism which had been implicitly assumed in physics since the seventeenth century. As a result, the purpose of physics became, not the discovery of laws enabling future physical events to be predicted with exactitude, but rather the discovery of laws enabling such events to be predicted *within the limits imposed by the uncertainty principle*.

The ineliminable presence of probabilities in quantum theory enables light to be shed on the question: how are the wave-like and particle-like aspects of quantum objects such as electrons and photons related? It was the Austrian physicist Erwin Schrödinger and the German physicist Max Born who, in the 1920s, provided the answer to this question. In 1925 Schrödinger introduced an abstract mathematical wave called a *wave function* which described the quantum state of a physical system. He formulated an equation—the celebrated *Schrödinger equation*—analogous to the well-known equations for physical wave motions, which describes the evolution of the wave function in time. The Schrödinger equation plays essentially the same role in quantum mechanics as does Newton's second law of motion in classical mechanics: both describe how a physical system changes over time.

In 1926 Born interpreted Schrödinger's wave function as a *probability wave* giving the probability that a particle or particles in a particular state will be measured to have a given position or momentum. Thus, through a remarkable synthesis of the Chance/Necessity and Continuous/Discrete oppositions, quantum theory invoked the concept of probability to unite waves and particles.

The Danish physicist Neils Bohr, one of the founding fathers of quantum theory, expanded wave-particle duality into a full-blown metaphysical principle—the *Principle of Complementarity*. This principle asserts that properties of physical objects occur in pairs. These, which Bohr called *complementary* pairs, bear a certain resemblance to opposed concepts. Each property in a complementary pair—for example, the position and momentum of an electron, or the wave and particle aspects of a photon—can be measured, but as the precision of the measurement of the one increases, that of the other must necessarily decrease. In the extreme case, if the measurement of one of the properties could be made perfectly precise, *no information whatsoever* can be obtained concerning the other property. That is, in physical terms, the other property would become *undefined*. Bohr concluded that quantum objects cannot be regarded as possessing well-defined inherent properties which are independent of their determination by means of a measuring device. According to Bohr, the quantum world does not possess a reality independent of the world of everyday experience which encompasses the measuring devices and experimental apparatuses used to probe the quantum world. On the other hand, the very existence of the world of everyday experience, built as it is from quantum objects such as electrons and photons, is dependent on the quantum world. The quantum world and the everyday world coexist in a state of mutual dependence, and, in Bohr's view, they constitute complementary aspects of an indivisible whole.

QUANTUM TUNNELING

It has long been observed that the wave-like character of light gives it the remarkable ability to *jump gaps*. For example, when a beam of light enters a block of glass at a shallow angle, it becomes trapped within the glass. (It is, of course, not blocked by the "barrier" air at the far side, but it is totally reflected back into the layer of glass, due to the lower refractive index of the air.) But if a second glass block is placed close to the first one, without touching it, then the dispersed nature of the light wave enables part of it to pass through the air barrier, and then penetrate and continue to travel through the second glass block. In this way the light beam apparently jumps the air gap and escapes from its trap.

A similar phenomenon occurs at the subatomic level in radioactive decay, in which an alpha particle tries to escape an unstable nucleus. The alpha particle is effectively confined in the nucleus by nuclear forces and, in principle, should not be able to escape. Nevertheless, it does manage to do so, by means of a process known as *quantum tunneling*.

We have seen that in quantum theory the behaviour of quantum objects can be described by means of probability waves. Because of this, quantum theory predicts that there is a nonzero probability that an object trapped behind a barrier, lacking the energy to pass through it, may occasionally appear on the other side of the barrier, without actually passing through it or breaking it down. This is quantum tunneling.

Quantum tunneling can be explained by invoking the uncertainty principle. For example, an alpha particle trapped within the nucleus of an unstable atom is highly localized in space, and its position can be pinned down with great accuracy. In that case, according to the uncertainty principle, its velocity must be indefinite, possibly much greater than would be expected. In particular it could move fast enough to escape the pull of the nucleus.

THE RIDDLE OF POLARIZATION

In the classical wave theory of light, a beam of light in which individual waves are aligned parallel to one another is said to be *polarized*. The common direction in which the waves of light are aligned is called the *direction of polarization* of the beam. Polarized light can be produced in a number of ways, for example by reflection from a transparent material, such as glass, or by passage through certain crystals or Polaroid plastic. A material that polarizes light is called a *polarizer*. The direction of polarization of the light issuing from a polarizer is called its *axis of polarization*.

Now suppose that light passes through one polarizer *A*, and then through a second polarizer, *B*. The beam of light issuing from *A* will be polarized along *A*'s axis of polarization, and after passing through *B* the intensity of the beam will be, in general, diminished, and polarized along *B*'s axis of polarization. The beam will emerge with undiminished intensity from *B* only if the latter's axis of polarization is parallel to that of *A*. If *B* is rotated, the intensity of the light issuing from it diminishes continuously until, when its axis of polarization is at right angles to that of *A*, no light issues from it.

In effect, the polarizer acts not as an apparatus "measuring" a photon's polarization before it enters the polarizer, but rather as a device which "prepares" the polarization of photons passing through it. The polarizer ensures that the polarization of photons emerging from it is always aligned along *its* axis of polarization. After a photon succeeds in passing through a vertical polarizer, it is *certain* to pass through a second vertical polarizer, thereby showing it to have vertical polarization. In this case it seems reasonable to say that the photon has vertical polarization *immediately before* the photon passes the second polarizer, since it was "prepared" by the first polarizer to have vertical polarization. On the other hand, if an "unprepared" photon impinges on the second polarizer, there is no more than a certain probability that the photon will pass through it. If it does, the emerging photon has vertical polarization *whatever* its polarization was before it entered the polarizer. We conclude from this that, except when the polarization state is "prepared," as above, *before its polarization is "measured" a photon has no determinate polarization.* It is, so to speak, the measurement itself which "causes" the photon's polarization to become determinate. The fact that the "effect" of the polarizer on the photon cannot, even in principle, be predicted with certainty shows that quantum theory is, unlike classical physics, *indeterministic.* In quantum theory Chance, rather than Necessity, rules.

Suppose now that we are given three polarizers *A*, *B*, and *C*. *A* and *B* are vertically and horizontally polarized, respectively, and *C* is obliquely polarized, specifically, *diagonally* polarized at 45°. Now if a photon succeeds in passing through *A*, it acquires vertical polarization, and we can be certain *both* that it will pass through a second copy of *A*, *and* that it will fail to pass through *B*. But if the photon succeeds in passing through *C*, thereby acquiring diagonal polarization, we can only assert that there is a *definite probability* (in this case, 1/2) that it will subsequently pass through *A* or *B*. According to quantum theory, there is no way of predicting with certainty whether the photon, having passed through *C*, will succeed in passing through *A* or *B*. This is an instance of *quantum indeterminacy.* In quantum theory, the result of an experiment is, in general, not—as would be the case in classical physics—fully determined by the conditions under the experimenter's control. The most that can be predicted is a range of possible results, together with a probability of occurrence for each one.

A vivid illustration is provided by the so-called *three polarizer phenomenon.* If we line up *A* and *B* in the order *AB* (or *BA*), no light will pass through, since the axes of polarization of *A* and *B* are at right angles. If we place *C* in front or behind *AB*, that is, in the order *CAB* or *ABC*, again no light will get through. But if we insert *C between A* and *B*, that is, if we line up the polarizers in the order *ACB*, some light *does* get through.

Consider what happens to an individual photon as it encounters the polarizers. If it passes through *A*, it emerges with vertical polarization; if it passes through *B*, it emerges with horizontal polarization; if it passes through *C*, it emerges with diagonal polarization. Supposing that the photon succeeds in passing through *C*, it has a 50% chance of subsequently passing through *A*, but no chance of then passing through *B* (since by passing through *A* it will have acquired vertical polarization which would prevent it from passing through *B*). So the photon has no chance of passing through *CAB*. On the other hand, if it passes through *A*, it then has a 50% chance of passing through C, and a further 50% chance of passing through *B*. So once a photon successfully passes through *A* it has a 25% chance of passing through *CB*. Thus, in general, a photon has a nonzero chance of passing through *ACB*.

It makes little sense to ask what the polarization of the photon was *before* it enters *ACB* or *CAB* (unless, of course, the photon has previously been 'prepared' to have a certain polarization by successfully passing through a polarizer with a known polarization). Whatever that polarization "was," all one knows is that the photon will *never* pass through *CAB* but that there is a nonzero probability that it will succeed in traversing *ACB*, emerging with horizontal polarization. And even if the photon has been "prepared" for vertical polarization by passing *A*, if it then successfully passes *CB* it emerges with horizontal polarization and so has *no chance* of passing *A* again.

The fact that an obliquely polarized photon has a certain probability of passing through a vertical polarizer and a horizontal polarizer indicates that, in some sense, the photon is partly in a state of vertical polarization and partly in a state of horizontal polarization. In quantum theory the polarization state of an obliquely polarized photon is called a *superposition* of vertical and horizontal polarization. Thus, instead of asserting that the polarization state of a photon before it meets a polarizer is merely indeterminate, we can now describe its state as a superposition of vertical and horizontal states. When the photon passes through a (vertical or horizontal) polarizer, its superposition state "collapses" into a state of vertical polarization or into a state of horizontal polarization. The photon's polarization state is compelled to make a sudden jump from being partly in each of these two states to being entirely in one or the other of them. Which of the two states it will jump into cannot be predicted with certainty, but the probability of each outcome can be calculated.

This discontinuous "collapse" of a quantum object in a superposition state into one of the superposed states is a cornerstone of the so-called *Copenhagen interpretation* of quantum mechanics put forward by Bohr and Heisenberg in the 1920s. The core principle of this doctrine is that quantum theory does not furnish a description of an objectively existing physical world but is concerned only with the probabilities associated with the observation or measurement of quantum objects, entities which can be in superposed states and so conform neither to the classical idea of particles nor the classical idea of waves. The act of measurement causes the range of possible values associated with the superposition to collapse, instantly and randomly, to exactly one of these possible values.

SCHRÖDINGER'S CAT PARADOX

Einstein and Erwin Schrödinger, another major figure in the early development of quantum theory, attacked the idea of superposition and, more generally, the Copenhagen interpretation by arguing that it would also have to apply to everyday objects, leading to paradoxical results.

Schrödinger's cat paradox, devised in 1935, provides an example of this. A simple version of the paradox can be set up using a so-called *horizontal-vertical (HV) polarizer*. Such a polarizer (e.g., a calcite crystal) is able to separate light impinging on it into two components of perpendicular polarization—horizontal and vertical. Unlike ordinary polarizers which absorb some of the incident light, an HV polarizer allows all the light through, splitting it into two components which travel along different paths. We also require that the HV polarizer includes a detector which is connected to a meter with a pointer that can assume one of three possible positions: a position O corresponding to the initial state before a photon passes through the polarizer, and positions H and V corresponding to a photon having passed through the horizontal and vertical channels respectively. Thus as a diagonally polarized photon passes through the HV polarizer, there is a 50% chance of the pointer reading H, and a 50% chance of it reading V. Before it passes through the polarizer, the state of the photon is a superposition of the vertical and horizontal states.

In Schrödinger's somewhat macabre scenario, inside a sealed box have been placed a light source, an HV polarizer with detector, a flask of hydrocyanic acid, a small hammer, and a cat. The point of the detector is connected to the hammer in such a way that if a vertically polarized photon is detected, the hammer shatters the flask, releasing the hydrocyanic acid, and the unfortunate cat is killed. If a horizontally polarized photon is detected, the hammer is unaffected and the cat remains alive. Now what happens when a diagonally polarized photon is emitted by the light source and passes through the HV polarizer? If the cat is regarded simply as a measuring apparatus, the answer is clear: the cat dies if the photon is detected as vertically polarized, and remains alive if it is detected as horizontally polarized. But suppose that an observer outside the box who accepts the Copenhagen interpretation is asked to report on the state of the cat. He cannot draw any conclusion about the state of the system (in particular, the cat) until it has been measured, which, from his point of view, only takes place when the box is opened and the state of health of the cat (alive or dead) observed. Before the photon passes through the HV polarizer, its polarization state is, as we have noted, a superposition of vertical and horizontal polarizations. Analogously, from the observer's standpoint, before the box is opened, the cat's state of health is a superposition of the states of being alive and being dead. According to the Copenhagen interpretation, by opening the box, and observing the cat, the observer "collapses" the cat's alive/dead superposition state into one of the two states, alive or dead. The observer cannot perceive a superposition of the two states.

There is a variant of Schrödinger's cat paradox known as *Wigner's friend* (after E.P. Wigner, another major figure in the development of quantum theory). This variant brings into relief the distinction between the conscious observer and a conventional measuring apparatus. In this scenario, the cat is replaced by a human "friend" of the observer's (in this case, Wigner) and the flask and hammer dispensed with altogether. When Wigner opens the box, he asks his friend what has taken place, and she will tell him that the pointer moved to H or V at a certain time. Assuming that the friend is truthful, the whole box and its contents cannot, as in the cat paradox, be treated as a quantum system since the friend would then have to be in a state in which she does not know whether the pointer was at H or V until she is actually asked by Wigner himself. This seems absurd, for while the cat may have been in a superposition state, surely the state of the friend's mind is quite certain—to her if to nobody else.

This brings out the sheer oddness, the paradoxical nature, of quantum theory when it is applied to objects of ordinary size. It is one thing to say of a microscopic quantum object that it is in a superposition of states, so that, for example, a photon can be half vertically and half horizontally polarized. We can just throw up our hands and resign ourselves to the fact that the quantum world is bizarre and quite different from the world of everyday objects. But when quantum theory is extended to the everyday world, this bizarreness cannot be avoided. Before the box is opened, is the cat *really* alive or dead? What if, for example, the pointer moves to 12 p.m. but the observer doesn't open the box until 1 p.m.? Common sense would dictate that the cat died at 12 p.m. and remained dead thereafter. But the Copenhagen interpretation decrees that, from the observer's viewpoint, between 12 and 1 p.m. the cat was in a superposition state, half-alive and half-dead, until 1 p.m. when it was observed to be dead. This leaves us with the question of *exactly when* quantum superposition ends and collapses into a definite state.

INTERPRETATIONS OF QUANTUM THEORY

A number of alternative interpretations of quantum theory have been proposed to answer the questions raised by Schrödinger's cat paradox, and others like it. The *statistical interpretation* posits that quantum-mechanical properties such as superpositions do not apply to individual systems—in particular to a single particle—but are purely mathematical, statistical quantities which are applicable only to an *ensemble* of similarly prepared systems. Under the statistical interpretation, the quantum-mechanical description would not apply to individual cat experiments, but only to the statistics of many similarly prepared experiments. By discarding the notion that quantum theory provides descriptions of individual physical systems, the statistical interpretation essentially trivializes quantum-theoretical paradoxes such as Schrödinger's cat.

In the so-called *objective collapse theories*, superpositions collapse spontaneously—independently of external observation—when some definite physical threshold (e.g., of time, mass, irreversibility, etc.) is attained. Given this, the cat will have assumed a definite state some time before the box is opened. Objective collapse theories demand that standard quantum theory be modified so as to admit what is known as *decoherence*, that is, to allow superpositions to collapse spontaneously after a certain amount of time has passed.

In the *relational interpretation*, no fundamental distinction is made between the human experimenter, the cat, or the apparatus, or between animate and inanimate systems. They are all considered quantum systems governed by the same quantum-theoretical rules; in particular, all may be considered "observers." But, as in the famous movie *Rashomon*, the relational interpretation grants the possibility that different observers may offer different accounts of the same series of events, depending on the information each possesses. The cat may be regarded as an "observer" of the apparatus, while the experimenter may be treated as another observer of the system in the box (the cat plus the apparatus). Before the box is opened, the *cat*, through the fact that it is either alive or dead, carries information about the state of the apparatus—namely, that the photon is horizontally polarized or vertically polarized. But at this stage the *experimenter* has no information concerning the state of the box's contents. Before the box is opened, the state of its contents would differ according to whether one takes the cat's, or the experimenter's standpoint. From the cat's standpoint, the apparatus is no longer in a superposition state—the latter has "collapsed" to a definite state. (If the cat is dead we may take the phrase "from the cat's standpoint" simply to mean "inside the box.") From the point of view of the experimenter, on the other hand, the box's contents appear to be in a superposition state. Not until the box is opened, and both observers have the same information concerning what has happened, do both system states appear to "collapse" into the same definite state, a cat which is either alive or dead.

The *many-worlds* (or "no-collapse") *interpretation*, devised by the American physicist Hugh Everett in 1957, draws the most radical conclusions from quantum theory about the nature of the physical world. Everett's idea was to eliminate the need for an observer in quantum theory. Instead, each time a (quantum) measurement is taken, the whole universe, including any would-be "observers," splits into a number of different "branches," or "worlds," each of which embodies a possible outcome of the event. For example, suppose Tom Swift throws a single die. Once thrown, there will be just one outcome—the die will show a definite number from 1 to 6, with an equal chance for each number. In the many-worlds interpretation, *all* six possible outcomes of throwing the die are realized, but each outcome is realized in a different "world" or branch of the universe—in this case, there are six branches Each of these contains a version of Tom (along with everything else) who sees an outcome of the throw of the die: in universe 1, Tom 1 who sees the die show a 1, in universe 2, Tom 2 with the die coming up 2, etc. None of the versions of Tom in the different universes is directly aware of the existence of the others. However, if any one of them is an adherent of the many-worlds interpretation, he will infer that they do indeed exist.

Now we have used an everyday event to illustrate the many-worlds interpretation, but in fact it was originally formulated to shed light on the nature of quantum-theoretic collapse of superposition states. In essence, the many-world interpretation simply *denies* that when a quantum measurement is made, any "collapse" of the superposed state occurs at all. Thus in the case of the HV polarizer, when a diagonally polarized photon is observed to pass through exactly one of the channels (H or V), this fact is explained in the Copenhagen interpretation as the result of a "collapse" of the photon's superposition state to a definite state. In contrast, the many-worlds interpretation denies that the measurement induces a "collapse" into a single state (H or V) but that in fact *both* outcomes (H and V) occur, only in *separate* universes.

In formulating the many-worlds interpretation, Everett's key insight was that, for a composite system—for example a subject (the "observer" or measuring apparatus) observing an object (the "observed" system, such as a particle)—the assertion that either the observer or the observed have separate well-defined quantum states is, strictly speaking, *meaningless*. Only the composite entity observer + observed has an absolute, well-defined state. To put it another way, in the process of observation, it is only possible to specify the state of the one *relative* to the other. Here we encounter another instance of the Absolute/Relative opposition. After the observation is made, the state of the observer and the observed become correlated or *entangled*. Each component of the superposition of the composite system observer + observed object contains two "relative states": a "collapsed" object state and an associated observer who has observed the same collapsed outcome. What the observer sees and the state of the object have become correlated by the act of measurement or observation. Thus, in the HV polarizer case above, the composite system is observer + pointer reading. After the observation this has "split" into the superposition of *observer seeing H + pointer reading H* with *observer seeing V and pointer reading V*. The subsequent evolution of each pair of relative observer–object states then proceeds in complete independence of the other components of the composite system, *as if* collapse has occurred. Since the superposition appears to have collapsed, Everett concluded that there was no need to assume that it had *actually* collapsed. And so he banished superposition collapse from the theory.

Although the many-worlds interpretation, with its proliferation of universes, may seem bizarre, it does have the merit of resolving many of the puzzles connected with quantum measurement.[1] For example, consider Schrödinger's cat. In place of a single cat in a puzzling half-alive and half-dead state, there are simply two cats, each in its own box in its own universe, along with two observers, each opening his own box and drawing his own conclusions concerning the state of the cat and so also of the photon. In the case of Wigner's friend, it is no longer necessary to be concerned about whether the measurement is made when his friend observes the pointer reading or when Wigner learns the result. As the photon passes through the polarizer it is split, along with Wigner, his friend, and the rest of the universe. In one universe the friend sees the pointer move to H and informs Wigner of this result. In the other she sees the pointer move to V and duly reports that result. The cat is both alive and dead—but it's not the same cat!

The difference between the many-worlds interpretation and the Copenhagen interpretation is strikingly illustrated by the idea of *quantum immortality*. This is based on a variation of the Schrödinger's cat thought experiment. In this version the cat is replaced by *the experimenter himself*, and the light source modifies so as to emit only diagonally polarized photons. The probability that the experimenter survives the first photon measurement is 50%, under both interpretations. But when subsequent photons are measured, a difference between the two interpretations appears. For under the Copenhagen interpretation, the quantum state of the system has already collapsed, so if the experimenter is already dead, there is a 0% chance of him surviving. But under the many-worlds interpretation, branches of the universe in which (copies of) the experimenter is still alive *always* exist, no matter how many photon measurements are made. If the

1 As we have seen, it can also help in resolving the conceptual difficulties of time travel.

experimenter's consciousness persists in these copies, then in a sense the experimenter could continue to live indefinitely.

The many-worlds interpretation also restores *determinism* to the quantum-theoretic picture of the universe. For consider, when I throw a die, I see only one outcome, 5 say, which I could not have predicted with certainty. But from the standpoint of the many-worlds interpretation, it is completely determined that, if the die is perfectly cubical, the universe will split into precisely six branches, each branch correlated with exactly one face of the die. The indeterminacy of my throwing a 5 is reduced just to the fact that I happen to occupy a particular branch of the larger ensemble of universes posited by the many-worlds interpretation, one in which I threw a 5. In that larger ensemble of universes, determinism is global, while indeterminacy is a local phenomenon arising from the fact that observers are confined to just one branch of the ensemble of universes at a time.

THE EPR PARADOX AND NONLOCALITY

In 1935 Einstein and his colleagues Boris Podolsky and Nathan Rosen showed that quantum theory had a further astounding consequence, namely, that spatially separated particles can influence each other *even when there is no apparent interaction between them*. This phenomenon, known as the *EPR paradox*, or *nonlocality*, is an instance of a still more radical departure of quantum from classical physics than those we have already discussed.

As originally formulated, the EPR paradox rested on the surprising insight that particles can interact in such a way that it becomes possible to measure both their position and their momentum more accurately than Heisenberg's uncertainty principle allows. Physicists have formulated an experimentally testable formulation of the paradox in the form of a physical system consisting of atoms in which a transition occurs from an excited state to the ground state accompanied by the emission of two photons in rapid succession. While the wavelengths of these photons are different, so that they correspond to different colours, red and blue, say, it is found that their polarizations *are always at right angles*. That is, if polarizers and detectors are set up to determine the polarizations of the two photons, it is invariably the case that, for example, if the red photon detector shows the red photon to be vertically polarized, then the blue photon detector will show the blue photon to be horizontally polarized.

But this renders the presence of one of the detectors *unnecessary*! For suppose we discard the blue photon detector. Then, since we know that the polarization of the blue photon is always perpendicular to that of the red photon, we need only measure the polarization of the red photon in order to determine the polarization of the blue photon. That is, the act of measuring the red photon's polarization also constitutes a measurement of the blue photon's polarization. Now we have already seen that before the measurement of a photon's polarization is actually made, its polarization is indeterminate. So if the red photon detector indicates that the photon has vertical polarization, what it is actually doing is "preparing" the photon's polarization state to be vertical. But once this has taken place, we *know* that the blue photon must have horizontal polarization—without in any way measuring or preparing it. The red polarizer has presumably prepared the red photon to be polarized V. But how could the red polarizer "cause" the blue photon to be polarized H without interacting with it? It would seem that the red polarizer must be "acting at a distance" on the blue photon. Moreover, the distance between the red polarizer and the blue photon at the time the measurement on the red photon is made can be arbitrarily large—even, in principle at least, light years. In this case the influence of the polarizer on the blue photon would have to be transmitted at speeds exceeding that of light, violating the special theory of relativity. This is the paradox.

There seem to be just two possible explications of the paradox. In the first explication quantum theory is retained, and the red polarizer genuinely "acts at a distance" on the blue polarizer. In essence, the ensemble consisting of the blue and red photons must be regarded as a *single, indivisible whole* in which the 'distance' between the photons—the parts of the whole—is irrelevant. This explication is what may be called *nonlocal*: it provides a vivid illustration of the Whole/Part opposition. Einstein and his coauthors rejected this explanation on the grounds that it violated the special theory of relativity. Instead they maintained that the outcome of the measurement on the blue photon is predetermined by some intrinsic property of the photon. Accepting this second, or *local* explication, entails abandoning the indeterminism of quantum theory and replacing it with what has become known as a *deterministic hidden variable theory*. The basic principle underlying such theories is that the state of a physical system, as formulated quantum-theoretically, fails to provide a complete description of the system; and that a complete theory would introduce new factors ("hidden variables") to account for all the observable behaviour of the system and so avoid indeterminism. Thus, in the case of the EPR paradox, a hidden variable theory would eliminate the nonlocality by showing that the outcome of the measurement on the blue photon is predetermined by some intrinsic property of the photon—the value of some "hidden variable"—unrecognized by quantum theory.

Einstein objected to the probabilistic and indeterministic nature of quantum theory, famously declaring "God does not play dice." By bringing out the nonlocal character of quantum theory, his paradox was designed to persuade his fellow physicists of its incompleteness, and so to search for a complete theory based on hidden variables. It has been shown that no hidden variable theory which avoids nonlocalities is compatible with the experimental results on state-correlated distant photons and particles which have been obtained in recent years. *Nonlocal* hidden variable theories compatible with these experimental results have been constructed, however. At present nonlocality seems to be unavoidable.

Chapter VII

Cosmic Enigmas[1]

THE BEGINNINGS OF COSMOLOGY

Cosmology, the study of the cosmos, the universe as a whole, is one of the grandest of intellectual games. It is an intoxicating mixture of physics, mathematics, philosophy, and theology, flavoured with a dash of sheer speculation. It is a source of puzzles and paradoxes. As the problem of ultimate origins, cosmology has gripped our thinking since the moment in the remote past our predecessors began to think in conceptual terms. Underlying cosmology are to be found many of the oppositions which have provided the underlying theme for this book: the Finite and the Infinite, the Whole and the Part, the One and the Many, the Constant and the Changing, Chance and Necessity.

1 This chapter is adapted from John L. Bell. "Cosmological Theories and the Question of the Existence of a Creator, "Chapter 6 of *Religion and the Challenges of Science*, edited by William Sweet and Richard Feist. Ashgate, 2007. Reprinted with the permission of Taylor and Franc is, UK.

Cosmology also provides a striking embodiment of the opposition between Being and Nothingness.

Engaged in cosmological thinking the mind seems to expand to encompass the universe. G.K. Chesterton put this memorably:

> The Cosmos is about the smallest hole that a man can hide his head in.

Cosmologists are the great dreamers, and so the great believers, in science. As the Russian physicist Lev Landau has observed:

> Cosmologists are often in error, but never in doubt.

Early cosmological theories chiefly took the form of *creation myths*. In the ancient Egyptian account of the creation of the cosmos, for example, the sun god Ra conjured up the universe from Nu, the swirling watery chaos:

> Heaven and Earth did not exist. And the things of the earth did not exist. I raised ... them out of Nu, from their passive state. I have made things out of that which I have already made, and they came from my mouth.

In the Vedas, the earliest Hindu scriptures, Reality or Being is described as having "arisen from Nothing." By contrast, in the cosmology of Jainism, the ancient Indian cosmology, time has no beginning. The universe was not created and has always existed. In Plato's *Timaeus* the universe is conceived as not having existed eternally, but as having been fashioned at some past time by a demiurge—a kind of universal craftsman outside of time—acting on formless matter already in existence.

The arresting first line of Genesis,

> In the beginning God created the heaven and the earth,

has customarily been taken to mean that nothing whatsoever (apart from the Deity itself) existed before the creation of the universe by God. This is a version of the doctrine of *creatio ex nihilo*, creation from nothing: it is one of the most fundamental instances of the opposition between Being and Nothingness. In the middle ages philosophers devised a number of logical arguments for *creatio ex nihilo*. The most prominent of these is the so-called *first cause*, or *cosmological argument*. This may be succinctly stated:

1. everything that begins to exist has a cause
2. the universe began to exist
3. therefore, the universe must have a cause

The cause of the universe inferred in 3 is then identified as God. Both premises of this argument can be—and have been—denied. The first premise will be falsified if closed causal loops exist as considered in Chapter IV, while the universe in both Jainist cosmology and steady-state cosmology (see below) has no beginning.

The first truly scientific cosmology was born with the publication of Isaac Newton's *Principia* in 1687. Resting on his theory of universal gravitation, Newton's universe was spatially infinite, and populated by infinitely many stars more or less evenly distributed in static gravitational equilibrium. Newton thought that any other stellar arrangement, for example the presence of only finitely many stars, or their uneven distribution, would cause them to clump together. Echoing Plato's account in the *Timaeus*, Newton held that space and time (and possibly matter as well) have always existed, but at some point in the past God acted to introduce order into the universe. In Newtonian cosmology the universe was essentially static: it was not conceived, on the cosmic scale, as possessing a history. While God intervened at some past moment to impose order on the universe, the precise time at which this miracle came to pass was held to be irrelevant, and so could be placed indefinitely far back in the past. In the Newtonian scheme, moreover, space and time themselves, were, like God itself, sempiternal, that is, of never-ending duration, and so not subject to the problem of origin. Thus the origin of the universe could be safely consigned to "minus infinity" and the attendant difficulties of identifying the time of that origin ignored.

But Newton's cosmology was soon seen to be vulnerable to the so-called *dark night sky paradox*, also known as *Olbers' paradox* (named after the eighteenth/nineteenth-century German astronomer Heinrich Wilhelm Olbers). This is the observation that the darkness of the night sky conflicts with the assumption of an infinite and eternal static universe, as postulated by Newton. For if the universe is static and populated by an infinite number of stars, then any line of sight from the earth must end at the (very bright) surface of a star, so the night sky should be uniformly bright, in contradiction with its evident darkness.

There are three possible resolutions of this paradox: the universe is (1) spatially finite, or (2) has a finite history, or (3) is not static. Cases (1) and (2) embody the Finite/Infinite opposition and (3) the Constant/Changing opposition. In case (1), few lines of sight from the earth meet the surface of a star, which is compatible with a dark night sky. In case (2) infinitely many stars may exist, but if they only sprang into existence at a finite time in the past, their light, travelling at a finite speed, has not had time to reach us yet. In case (3) the universe may be infinite in space and time, but if it is uniformly expanding at a speed which increases with distance, then starlight reaching the earth will be red-shifted so as to reduce its energy sufficiently to be in conformity with a dark night sky.

As we shall see, each of these possibilities has been taken up by cosmologists.

STEADY-STATE VS. BIG BANG

In 1917 Einstein applied the general theory of relativity to the problem of cosmic structure. He retained Newton's assumption of a static universe, in the first place because that was the way the universe still appeared to astronomers at the time and also because, as he remarked, "one would get into bottomless speculations if one departed from [that assumption]." But, like Newton, he was faced with the problem of stellar clumping. Worse, according to his theory a universe containing matter would contract under gravity and ultimately collapse altogether. To counteract this he introduced a force of "cosmic repulsion" into his mathematical model. (He later discarded this notion when it was established that the universe is actually expanding.) The Einstein universe is static, finite and like the surface of a sphere, unbounded.

In the 1920s the static view of the universe was overturned by two discoveries, one theoretical, the other empirical. In 1922 the Russian mathematician Alexander Friedmann, and, independently, the Belgian cleric Georges Lemaître in 1927, showed that Einstein's equations have solutions which represent a universe containing matter, but in a state of continual *expansion*. In 1929 Edwin Hubble, in apparent ignorance of Friedmann's and Lemaître's theoretical results, found the first evidence of such an expansion, observing a redshift in the spectrum of galaxies in direct proportion to their distance from us. Hubble's observations seemed at first to indicate that the expansion of the universe had begun only a billion or so years ago, contradicting the evidence from radioactivity in rocks that the earth's crust must be at least 5 billion years old (revised recently to 4.5 billion years). Happily, this conflict was resolved by a later revision of the distance yardstick based on stellar luminosity, which resulted in the origin of the expansion being pushed back to a point some 14 billion years in the past.

The discovery that the universe is expanding naturally led to the question of when and how the expansion began. Lemaître suggested that the universe had evolved by explosive expansion from an initial highly compressed and extremely hot state which he called the "primeval atom"—the first explicit formulation of what later came to be known as the "Big Bang."[2] The apparent support of science for the idea that the universe had an origin of some kind naturally appealed to cosmologists of a Christian persuasion, among whom E.T. Whittaker and E.A. Milne were prominent. Whittaker held that God created the universe *ex nihilo*:

> When the development of the system of the world is traced backwards by the light of laws of nature, we arrive finally at a moment when that development begins. This is the ultimate point of physical science, the farthest glimpse that we can obtain of the material universe by our natural faculties. There is no ground for supposing that matter ... existed before this in an inert condition, and was in some way galvanized into activity at a certain instant: for what could have determined this instant rather than all the other instants of past eternity? It is simpler to postulate a creation *ex nihilo*, an operation of divine will to constitute nature from nothingness.

In 1951 Pope Pius XII cited Whittaker's assertion of the consonance between the Christian tradition and the picture of the expanding universe as providing scientific evidence for the Catholic world view.

Cosmologists of an agnostic tendency were understandably alarmed that their discoveries could be used, plausibly or not, to underpin traditional Christian theology. In particular the "Big Bang"

---

2 This term was introduced—with derisory intent—by physicist Fred Hoyle in 1950; but, ironically, it came into general merely descriptive use.

scenario, through its postulation of a "beginning" to the existence of the universe, seemed to offer disturbing new support for the First Cause Argument. Unwilling to become embroiled in theological dispute, many physicists welcomed the formulation of the *steady-state* theory of the universe in 1948 by Hermann Bondi, Thomas Gold, and Hoyle. This provided an alternative to the "Big Bang" scenario according to which the universe apparently sprang explosively into being from nothing and then expanded, continually cooling and attenuating, into its present quiescent state. The steady-state theory, by contrast, denied that the universe is any cooler or less dense at present than it was in the past, and that any change in the large-scale structure of the universe has occurred over time. Most importantly from a philosophical standpoint it is denied that there was ever a time in the past at which the universe has not existed. These assertions were based on what Bondi and Gold termed the *Perfect Cosmological Principle*. This is an extension of the fundamental *Cosmological Principle*—accepted by virtually all cosmologists—which asserts that, on a sufficiently large scale, the properties of the universe are the same for all observers, wherever they happen to be situated, although these features may change over time. The absence of such an assumption would make observational cosmology virtually impossible, for without it we could infer very little about the structure of the universe as a whole on the basis of what we can observe from the tiny part of the universe occupied by our planet. Here we see the opposition between the Whole and the Part at work. Bondi and Gold's Perfect Cosmological Principle extends the Cosmological Principle to time as well, so asserting that, on a sufficiently large scale, the properties of the universe are the same for all observers, wherever and *whenever* they happen to be situated. According to the Perfect Cosmological Principle the universe looks the same everywhere (on a sufficiently large scale), and its appearance does not change with time.

At first sight, the Perfect Cosmological Principle conflicts with the observed expansion of the universe. The devisers of the steady-state theory were well aware of this. They saw that the only way of preserving a changeless universe in the presence of expansion was to postulate a continual uniform creation of matter at precisely the rate required to offset the attenuation brought about by the expansion. A daring hypothesis indeed!

A vivid illustration of the steady-state theory is provided by the "Flying through Space" screensaver for personal computers. Here is presented, on a dark background, a continuous flow of "stars" radiating outwards from the centre of the screen. This is intended to represent, as its name implies, what an observer would see as she moves rapidly through interstellar space, encountering along the way "stars" which are already "in existence." But equally the picture may be viewed from the standpoint of an observer supposed stationary. In that case the "stars" must be taken as continually springing into being and receding outwards. Despite the continual emergence of new "stars," the law of conservation of mass (or energy) is obeyed "locally" in the sense that the total number of such "stars" on the screen does not change with time. This provides a perfect illustration of the opposition between the Constant and the Changing.

In order to compensate for the universe's expansion the steady-state theory called for the "creation" of just one hydrogen atom per cubic centimetre of space every 10^{15} years. This is a phenomenon well below the limits of conceivable observation. Cosmologists anxious to steer clear of the theological quagmire caused by the Big Bang theory were more than willing to accept an infinitesimal amount of "continuous creation" as the price to be paid for once again thrusting the origin of the universe back to minus infinity, where they instinctively felt it belonged. This was unquestionably an important consideration for Hoyle, who in a 1950 radio broadcast stated:

> Some people have argued that continuous creation introduces a new assumption into science—and a very startling assumption at that. I do not agree that continuous creation is an additional assumption. It is certainly a new hypothesis, but it only replaces a hypothesis that lies concealed in the older theories, which assume ... that the whole of the matter in the universe was created in one big bang at a particular time in the remote past. On scientific grounds this big bang assumption is much the less palatable of the two. For it is an irrational process that cannot be described in scientific terms. Continuous creation, on the other hand, can be represented by mathematical equations whose consequences can be worked out and compared with observation. On philosophical grounds, too, I cannot see any good reason for preferring the big bang idea. Indeed, it seems to me in the philosophical sense to be a distinctly unsatisfactory notion, since it puts the basic assumption out of sight where it can be challenged by a direct appeal to observation.

Steven Weinberg, a later champion of the Big-Bang theory, put the matter bluntly:

> The steady state theory is philosophically the most attractive theory because it *least* resembles the account given in Genesis.

The steady-state theory may be likened to the Parmenadian view of the world as an unchanging whole, while the Big Bang scenario embodies the Heraclitean view that everything is in a state of flux. Here we see another instance of the opposition between the Constant and the Changing.

While the steady-state theory did not lack philosophical appeal, the observational evidence soon began to mount against it. To begin with, continuous creation required particles and corresponding antiparticles, such as electrons and positrons, to be produced at equal rates, which would lead to a symmetry between matter and antimatter. But the observed universe shows no such symmetry. Rather it shows a marked preponderance of one sort of matter over the other. Moreover, the discovery of quasi-stellar objects ("quasars") at great distances showed that the universe did, after all, change its appearance with time.

The *coup de grâce* was delivered to the steady-state picture in 1965 with the discovery of the "echo" of the Big Bang. In 1948 the physicists Ralph Alpher, George Gamow, and Robert Herman had predicted that if the Big Bang scenario of a hot and dense past were correct, then some evidence of that past must remain in the form of residual radiation cooled by the universe's expansion to a temperature only a few degrees above absolute zero. In 1965 Arno Penzias and Robert Wilson happened upon this radiation field while calibrating a sophisticated radio receiver designed for satellite tracking. The radiation had a temperature of three degrees absolute, almost exactly as predicted. Subsequent observations established that its spectrum carried the distinctive Planck signature of heat radiation. The steady-state theory provided no plausible means of explaining the presence of a pervasive radiation field with just these characteristics. The Big Bang theory received further confirmation over its rival from the successful prediction of the cosmic abundances of the light elements helium, deuterium, and lithium, all of which would be produced by nuclear reactions during the first three minutes of the expansion after the Bang. The steady-state theory was incapable of explaining the abundance of these elements.

By the mid-1960s most cosmologists regarded the steady-state theory as dead. Nevertheless, despite the mounting observational evidence in favour of the Big Bang theory, Hoyle himself, along with a few of his followers, steadfastly refused to embrace it, no doubt because of a distaste for its postulation of a temporal origin of the universe. Hoyle's stubbornness in this regard was satirized in a verse by Barbara Gamow, the wife of George Gamow, himself a champion of the Big Bang theory. Here Ryle is the British radio astronomer Martin Ryle, who, as a proponent of the Big Bang theory, engaged in extended debate, much of it acrimonious, with Hoyle throughout the 1950s and 1960s.

"Your years of toil,"
Said Ryle to Hoyle,
"Are wasted years, believe me.
The steady state
Is out of date.
Unless my eyes deceive me,
My telescope
Has dashed your hope;
Your tenets are refuted.
Let me be terse:
Our universe
Grows daily more diluted!"
Said Hoyle, "You quote
Lemaitre, I note,
And Gamow. Well, forget them!
That errant gang
And their Big Bang—
Why aid them, and abet them?
You see, my friend, it has no end
And there was no beginning,
As Bondi, Gold,
And I will hold
Until our hair is thinning!"

THE PROBLEM OF THE ORIGIN OF THE UNIVERSE

The Big Bang theory soon came to be accepted by the majority of physicists. The prominent cosmologist Martin Rees, writing in 1999, had this to say on the matter:

> The empirical support for a Big Bang ten to fifteen billion years ago is as compelling as the evidence that geologists offer on our Earth's history.... A few years ago, I already had ninety per cent confidence that there was indeed a Big Bang—that everything in our observable universe started as a compressed fireball, far hotter than the centre of the Sun. The case now is far stronger: dramatic advances in observations and experiments have brought the broad cosmic picture into sharp focus during the 1990s, and I would now raise my degree of certainty to ninety-nine per cent.

Mathematical support for the Big Bang scenario had been independently provided by the *Hawking-Penrose singularity theorems*. These demonstrate under certain seemingly plausible hypotheses within the general theory of relativity that the entire cosmos must have emerged in the past from a *universal singularity*, that is, a place where the laws of physics break down. This singularity marks the "beginning" of the universe in time. (The hypotheses underlying the Hawking-Penrose theorem are: (i) gravity is attractive and universal, (ii) the universe is now expanding and contains sufficient matter, (iii) there are no closed time-like lines, that is, time travel into the past is impossible.) Before this universal singularity sprang into being, neither space, time, matter, nor the laws governing them can be held to have existed. In fact no meaning can be attached to the phrase "before the singularity." The universe must be regarded as having materialized *ex nihilo*. But Hawking and Penrose did not follow the theologians in putting forward a reason as to *why* this singular event came to occur. All they asserted is that, under the specified hypotheses, the universe and the laws governing it cannot always have existed: they must have materialized at some moment in the past.

But the question of the universe's origin continued to prey on the minds of cosmologists. Reluctant to surrender this problem to the theologians and philosophers, cosmologists attempted to remove the sting of the first cause argument by treating the spontaneous emergence of the universe as a problem in *physics*. They hoped that a suitable synthesis of quantum theory and relativity would enable the derivation of the initial singularity to be blocked. In that case, the putative "beginning" or "creation *ex nihilo*" of the universe need no longer be identified with something as nebulous as a singularity, but could actually be invested with *physical content*. One such proposal was put forward (perhaps with tongue in cheek) in the early 1970s by the physicist Edward Tryon. He suggested that the universe may be nothing more than a monstrous "vacuum fluctuation" in the sense of quantum field theory. A typical example of a vacuum fluctuation is the occasional emergence of an electron, positron, and photon from a perfect vacuum. When this occurs, the three particles exist only for a brief time, and then annihilate each other, leaving no trace behind. Energy conservation is violated, but only for the brief particle lifetime permitted by the "time/energy" formulation of the Heisenberg uncertainty principle (see Chapter VI). This has the consequence that the smaller the net energy of the particles, the longer the fluctuation can last. In particular, if the net energy is zero, then the particle lifetime can be any value whatsoever, however large. Now the laws of physics place no limit on the scale of vacuum fluctuations. So if the universe is closed and has zero net energy (for this to be possible the universe's total "positive" mass energy must be balanced by its total "negative" gravitational potential energy) it could be itself the result of a vast fluctuation of the vacuum of some hyperspace—the "quantum void"—in which our own universe is embedded. By way of explanation for this remarkable occurrence, Tryon engagingly offered

> the modest proposal that our Universe is simply one of those things that happen from time to time.

In Tryon's scenario, the emergence of our universe is a random occurrence, a sudden precipitation from the quantum void. This cannot be regarded as a creation *ex nihilo* since the "void" from which the universe sprang is assumed already to be present. Like Newton's cosmos, the quantum void has always been "there." The presence of the quantum void amounts to what might be called an "escape into infinity."

A far more radical proposal for avoiding the initial singularity was offered by James Hartle and Stephen Hawking in the early 1980s. In this—the so-called "no boundary" scenario—the universe's initial state is timeless, in that it possesses, not three spatial and one temporal dimension, but *four spatial dimensions*, the additional spatial dimension being called by Hawking "imaginary time." (The idea of introducing "imaginary time" was first proposed by Minkowski in 1908 in order to allow the metric of special relativity to assume a Euclidean form.)

In the "no-boundary" scenario the spacetime geometry of the initial state assumes a Euclidean form, which makes it possible for the universe to lack a "beginning" yet still be temporally closed. For just as the Earth's surface has no boundary at the North Pole, this initial region of the universe also lacks a boundary: it has no singular points. The geometry of the "no-boundary" universe is similar to that of the surface of a sphere, except that it has four dimensions instead of two. In this analogy, unfolding in Hawking's "imaginary time," Earth's North Pole represents the Big Bang, which, like the North Pole, is not a singularity. In that case, the universe itself cannot be said to have a "beginning" in the ordinary sense of the word, at least not in respect of imaginary time. Hawking has made strong claims for the objective existence of the latter:

> Only if we could picture the universe in terms of imaginary time would there be no singularities.... When one goes back to the real time in which we live, however, there will still appear to be singularities. This might suggest that the so-called imaginary time is actually the real time, and that what we call real time is just a figment of our imaginations. In real time, the universe has a beginning and an end at singularities that form a boundary to space-time and at which laws of science break down. But in imaginary time, there are no singularities or boundaries. So maybe what we call imaginary time is really more basic, and what we call real is just an idea that we invent to help us describe what we think the universe is like.

And on one occasion Hawking asserted:

> I still believe the universe has a beginning in real time, at the big bang. But there's another kind of time, imaginary time, at right angles to real

> time, in which the universe has no beginning or end. This would mean that the way the universe began would be determined by the laws of physics. One wouldn't have to say that God chose to set the universe going in some arbitrary way that we couldn't understand. It says nothing about whether or not God exists—just that he isn't arbitrary.

In the "no-boundary" scenario, the universe, viewed in imaginary time, is a kind of *ouroboros*, a snake eating its own tail. In view of this one might think that the "no-boundary" scenario avoids the singularity problem without making an "escape into infinity." But a closer look dispels this impression. For the "no-boundary" scenario rests upon an esoteric application of quantum theory to the collection of possible geometries of spacetime. This collection must already be assumed to be present, in an ontological sense, at least, prior to the actual universe. Like Tryon's "quantum void," it was always present. This is just another "escape into infinity," if a highly unusual one.

The "escape into infinity" is also to be seen in the so-called ekpyrotic model of the universe. This term, which derives from the Greek word *ekpyrosis*, "conflagration," is intended to evoke the ancient cosmological model associated with Heraclitus and the Stoics, according to which the universe is created (and recreated) in a sudden burst of fire. Here the hot Big Bang universe is conceived as arising from the collision of two three-dimensional worlds, or "membranes," in a five-dimensional space. The all-embracing five-dimensional space is, again, taken to be infinite and as having in some sense always "existed."

Although it seems fair to say that the majority of cosmologists avoid using the Big Bang to draw theological conclusions, by no means all are opposed to the notion that the universe was "created." John Polkinghorne, who resigned his professorship of physics at Cambridge to become an Anglican priest, is one example. The cosmologist Paul Davies's recent work gives evidence of an emerging deism. Some have argued that, given the Big Bang, the hypothesis that the universe was created has at least the merit of simplicity, and is therefore to be preferred to arcane conceptions such as Hawking's. This is the position espoused by the Catholic physicist Nicola Dallaporta, who has asserted:

> In order to justify the various ... assumptions current in present day cosmology, it is necessary for each of them to build a frame of

> metaphysical postulates much more involved and artificial than the opposite straightforwardly metaphysical view of a universe built according to an a priori plan, requiring a planning Intelligence adequate to having conceived it.

The real issue here is that the existence of a "planning Intelligence," while straightforward, is not merely a *metaphysical* claim, but a *theological* one. In understanding the universe, most cosmologists prefer a complex metaphysics to a simple theology.

DARK MATTER, DARK ENERGY, AND COSMIC ACCELERATION

Astrophysicists have long been perplexed by the observation that galaxies and galactic clusters rotate so quickly that the gravity produced by the observable matter they contain is far below what is required to hold them together. They should have flown apart long ago. This has led to speculation that some material yet to be detected directly is lending these galaxies extra mass, so furnishing the extra gravity they require to remain intact. This strange material, being invisible, has been termed "dark matter."

The existence of dark matter has been inferred solely from the gravitational effect it is presumed to have on visible matter. Unlike normal matter, dark matter does not absorb, reflect or emit light (or any other electromagnetic radiation), and so is extremely difficult to detect. Nevertheless dark matter seems to outweigh visible matter roughly six to one, constituting about 27% of the mass-energy content of the observable universe (surprisingly, "ordinary" matter provides just 5% of that content). Physicists are inclined to the view that dark matter is most probably composed of massive subatomic particles which interact only through gravity and the weak nuclear force. The search for this particle is one of the major programs in particle physics today.

Another perplexing observation, first made in the 1990s is that the *rate of expansion of the universe is increasing*, a phenomenon known as *cosmic acceleration*. In response, cosmologists have hypothesized the presence of an invisible substance permeating the universe which exerts a strong negative pressure on it. This—the so-called *dark energy*—would have the effect of accelerating the universe's expansion. The effect has been called, somewhat misleadingly, "gravitational repulsion." But in fact the negative pressure of dark energy has no influence on the gravitational interaction between masses, which remains attractive. Rather, it affects the rate of expansion of the universe as a whole. It is estimated that dark energy makes up 68% of the mass-energy content of the observable universe.

THE ARGUMENT FROM DESIGN VS. THE MULTIVERSE

If the attempt to explicate what, if anything, happened "before" the Big Bang has rekindled the first cause argument, then explaining what happened afterwards has ensnared cosmologists in a different, and perhaps better-known argument for the existence of a Creator. This is the so-called *argument from design*. The argument, which received its most celebrated elaboration in William Paley's *Natural Theology* of 1802, has been encapsulated by Bertrand Russell as follows:

> Everything in the world is made just so that we can manage to live in the world, and if the world was ever so little different, we could not manage to live in it.

And indeed, recent work has demonstrated the exactness of the "fine tuning" of the fundamental constants of nature necessary for ensuring that the universe has the requisite form making possible the formation of physical structures—galaxies, stars, planets—from which organic life, in particular ourselves, can eventually emerge. In his book, *Just Six Numbers*, the British cosmologist Martin Rees identifies these constants. They are:

- N, about 10^{36}, the ratio of the electrical to the gravitational force;
- E, about 0.007, which measures the efficiency of thermonuclear fusion of hydrogen to helium;
- Ω, the ratio of the actual density of matter in the universe to the "critical" density at which the universe will eventually recollapse;

- λ, the "cosmological constant"—originally introduced by Einstein—the ratio of the putative force of cosmic repulsion to the force of gravity;

- Q, of the order of 10^{-5}, the ratio of the energy required for complete dispersal of a galaxy or galactic cluster to its rest mass energy;

- D, exactly 3, the number of spatial dimensions of the universe.

Had the values of these constants differed even slightly from their actual values, the structure of the universe would be altered to such a degree as to make the emergence of any form of organic life impossible. In Rees's words:

> if *N* had a few less zeros, only a short-lived miniature universe could exist: no creatures could grow larger than insects, and there would be no time for biological evolution.
>
> *E* ... controls the power from the Sun and, more sensitively, how stars transmute hydrogen into all the atoms of the periodic table. Carbon and oxygen are common, whereas gold and uranium are rare, because of what happens in the stars. If *E* were 0.006 or 0.008, we would not exist.
>
> Ω tells us the relative importance of gravity and expansion energy in the universe. If this ratio were too high relative to a particular "critical" value, the universe would have collapsed long ago; had it been too low, no galaxies or stars would have formed.
>
> Fortunately for us ... λ is very small. Otherwise its effect would have stopped galaxies and stars from forming, and cosmic evolution would have been stifled before it could even begin.
>
> If *Q* were [any] smaller, the universe would be inert and structureless; if *Q* were much larger it would be a violent place, in which no stars or solar systems could survive, dominated by vast black holes.
>
> Life couldn't exist if *D* were two or four.

And, even if human beings could exist under the last of these conditions, they would find frustrating the fact that they couldn't tie their shoelaces, for there are no knots in even-dimensional spaces!

For the universe to have the structure it has, and in particular to have a structure compatible with the existence of life, it would seem, then, that these six numbers must have been "fine-tuned" to their actual values. As Rees points out, one could follow some scientists in responding to this with a shrug of the shoulders and the remark that since we couldn't exist if these numbers failed to have these special values, and since we manifestly *do* exist, there's nothing to be surprised about. This is a version of the so-called *weak anthropic principle*, which the cosmologists John Barrow and Frank Tipler define as follows:

> The observed values of all physical and cosmological quantities are not equally probable but ... take on values restricted by the requirement that there exist sites where carbon-based life can evolve and by the requirement that the Universe be old enough for it already to have done so.

Such a response is less than satisfactory in that it fails to provide an *explanation* for the remarkable fact that these six constants take just the values they do. The probability of their having done so would seem to be vanishingly small! This has led John Polkinghorne to suggest, in a revival of the argument from design, that the "fine-tuning" of these numbers furnishes genuine evidence for the intervention of a beneficent Creator, who formed the universe with the specific intention of creating organic life, and, more particularly, us.

Now the force of this contemporary version of the argument from design depends not only on the fact that the six constants have providential values, but also on the assumption that the universe in which they take these values is *unique*. The *Parable of the Frogs* illustrates this. The frogs in a certain pond, known to them as the *Universal Pond*, lead an idyllic life. Conditions in the Universal Pond are, the frogs note, perfectly adapted for froggy existence. The water temperature is just right, neither too hot nor too cold. On the surface of the Pond float a number of lily pads ideally designed, so think the frogs, for perching on. And there is an abundance of tasty insects providing an ideal source of nourishment. "Now surely," one can imagine a philosophically-minded frog arguing, "the conditions in the Universal Pond cannot have arisen merely by chance. For the Universal Pond is, by definition, unique, and it is far too unlikely that in this single case the temperature of the water, the dimensions of the lily-pads, and the constitution of the insects would have the ideal values they in fact possess.

The only reasonable explanation is that these conditions must have been brought about by an intelligent Designer." But what the frogs do not know is that in fact their Universal Pond is merely one among a vast ensemble of ponds, in which are manifested every conceivable variation of conditions. Many contain no frogs at all because the water is polluted or has dried up altogether. Others support a population of frogs leading a wretched flipper-to-mouth existence. The explanation for the pleasant ambience afforded by the "Universal" pond is in fact quite prosaic, amounting to no more, sadly, than the chance fact that its inhabitants happen to reside in a pond in which such agreeable conditions obtain. The fable might conclude with a frog of unusual acuity grasping the possibility that other ponds could exist, and enunciating the "weak froggy principle," namely: *The observed values of all physical and pondal quantities are not equally probable but ... take on values restricted by the requirement that there exist sites where carbon-based, and, more especially, frog life can evolve and by the requirement that the Universal Pond be old enough for it already to have done so.*

The Parable of the Frogs provides a vivid instance of the Chance/Necessity opposition. For the frogs, the ideal state of their Universal Pond is a Necessity. But in actuality it is just a product of Chance.

Concerned to avoid the conclusion that the universe was designed by a Creator, cosmologists have challenged the assumption of our own universe's uniqueness, suggesting that, instead, the universe we inhabit is just one among many, an element of a larger structure which has come to be called the *multiverse*. If cosmologists met the first cause argument by an "escape into infinity," their riposte to the argument from design is accordingly what might be termed an "escape into plurality." To quote Martin Rees again,

> If one doesn't accept the "providence" argument, there is another perspective, which—though still conjectural—I find compellingly attractive. It is that our Big Bang may not have been the only one. Separate universes may have cooled down differently, ending up governed by different laws and defined by different numbers. This may not seem an "economical" hypothesis—indeed, nothing might seem more extravagant than invoking multiple universes—but it is a natural deduction from some (albeit speculative) theories, and opens up a new vision of our universe as just one "atom" selected from an infinite multiverse.

> If indeed there were an ensemble of universes, described by different "cosmic numbers," then we would find ourselves in one of the small and atypical subsets where the six numbers permitted cosmic evolution. The seemingly "designed" features of our universe shouldn't surprise us, any more than we are surprised at our particular location within our universe.

The "multiverse" conception does not arise from science fiction, nor is it a direct result of the natural dialectical move from the One to the Many (although that surely provides its ultimate underpinning). Actually the multiverse conception emerged from the *inflationary universe* model devised by the cosmologist Alan Guth and others, which was originally introduced to explain the cosmological "horizon" and "flatness" problems. The "horizon" problem is the puzzle that widely separated regions of the universe are observed to share the same physical properties, such as temperature, even though these regions were too far apart when they emitted their radiation to have exchanged heat and so homogenized during the time since the Big Bang. The "flatness" problem is the question of why the universe today is so close to the boundary between being open or closed, that is, why it is almost "flat." The essential feature of the inflationary universe model is that, shortly after the Big Bang, the infant universe underwent a brief (perhaps as little as 10^{-32} sec.) and extremely rapid expansion, after which it resumed the more leisurely rate of expansion of the standard Big Bang model. The cosmologist Andre Linde has taken this idea further in formulating what he called *chaotic inflationary universe* models. In such scenarios, an inflating universe fissions into a number of different fragments or "bubbles," each of which is completely cut off from its fellows. These fragments are, in effect, independent "pocket" universes. This fissioning process thus turns the inflationary universe into a "multiverse." The process may be repeated at random in the new "universes," each of which accordingly spawns a whole flotilla of offspring or "baby" universes. These, in their turn, reproduce in the manner of their progenitors. Individual universes would come and go, but in Linde's vision the whole ensemble of universes—the "multiverse"—would last forever. Under the so-called eternal inflation scenario, these universes would continue to fission limitlessly, producing an unbounded number of pocket universes. Taken together these comprise a multiverse more than roomy enough to deal with the design problem. Unfortunately, however, this multiverse conception is still vulnerable to the first cause argument, since, in the words of Alan Guth, "while the issue is not completely settled, it appears likely that eternally inflating universes must necessarily have a beginning." By this he means that, given certain plausible physical assumptions, such universes are subject to some version of the Hawking-Penrose singularity theorems and so must contain an initial singularity—an "origin." In fact, Guth, in collaboration with the cosmologists Arvind Borde and Alexander Vilenkin, have proved a theorem which asserts that, in any expanding cosmology, including inflationary cosmologies, the universe must have had a beginning in the past. In particular the whole multiverse produced by eternal inflation, while expanding forever into the future, must itself have had a "beginning." To avoid this, Vilenkin has proposed that the universe "created itself" through a process of quantum tunneling.

A PHILOSOPHICAL CODA

Towards the end of *Just Six Numbers* Martin Rees remarks:

> If the underlying laws determine all the key numbers uniquely, so that no other universe is mathematically consistent with those laws, then we would have to accept that the "tuning" was a brute fact, or providence.

Let us suppose that the laws of physics *did* indeed ultimately turn out to determine all the "cosmic numbers" uniquely. Then the "tuning" of the fundamental constants would be a *logical* consequence of the laws of physics. In that case, would not the acceptance of the "tuning" as a brute fact, or providence, be tantamount to an acknowledgement that the *laws of physics* were *themselves* brute facts, or providence? Compare this with the analogous situation in mathematics. The value of π (say) is uniquely determined by the laws of mathematics, indeed, some would claim, by the laws of logic itself. But few are inclined to regard any such fact as "brute" or providential, because mathematics, or logic, is taken to have an *a priori* character, a character customarily denied to physics. Most theists do in fact accept that the deity is constrained to act in accordance with the laws of logic or mathematics—that is, if, as Plato said, God always geometrizes, it's because he has no choice but to do so! Such laws are acknowledged by theists to be "brute facts," only of a necessary nature over which God himself has no control. But if, by contrast, the laws of physics should turn out also to be unique or "brute facts," that very contingency would make it natural for the theist to claim that they had been expressly selected from the spectrum of possibilities by divine choice. In order to block this new meta-version of the argument from design without at the same time turning away from the goal of explaining the apparent uniqueness of the laws of physics, the agnostic, now with back against the wall, might be forced to take refuge, in a sort of "metaphysical pluralism," which countenances the *actual* existence of realms of being, inaccessible from ours,[3] and governed by entirely different physical laws. Faced with this possibility, the agnostic would have good reason to hope that the exact values of the "key numbers" of our universe are in principle undeducible, or at least will remain undeduced, from the laws of physics.

Despite the dramatic successes since the work of Einstein and Hubble which have led to our present view of the universe and its history, we are still a long way from a complete cosmology. Creation ex nihilo or eternality? Universe or multiverse? These and other questions, each resting on a fundamental and enduring opposition, still await definitive answers.

3 This is the view of David Lewis, expounded in his book *On the Plurality of Worlds* (Oxford: Blackwell, 1986). Lewis's view, however, derives from purely metaphysical considerations and has little to do with scientific cosmology.

Appendix 1

Paradoxes in Logic and Language

Here we give a brief account of paradoxes of a logical and linguistic nature.

THE LIAR PARADOX

The liar paradox purports to show that common beliefs about *truth* and *falsity* actually lead to *contradiction*. This is done by formulating perfectly grammatical statements which cannot consistently be assigned a truth value.

In its simplest form, the liar paradox is the statement: *the statement I am now making is false*, or, *this statement is false*. Since this statement asserts its own falsehood, if it is true, it is false, and if it is false it is true. So it is both self-referential and self-contradictory.

This argument depends on the *rule of double negation*, namely, the assertion that, for any statement *p*, if *p is false* is false, then *p* is true. This in turn is a consequence of the *law of bivalence* or *the law of excluded middle*, namely, that for any statement *p*, *p* is either true or false (but not both).

In fact the rule of double negation is not needed to derive the contradiction in the liar paradox. For let ℓ denote the liar statement. Then ℓ asserts *ℓ is false*. So if ℓ is true, ℓ is false, and it follows that ℓ implies not ℓ. Hence ℓ implies ℓ-and-not-ℓ, so that ℓ implies a contradiction. It follows that ℓ is false, i.e., *ℓ is false* is true. But *ℓ is false* is just ℓ, so it follows that ℓ is true. So ℓ is both true and false, a contradiction.

The liar paradox can be traced back as far as ancient Greece. The earliest known example of a liar-type statement is the *Epimenides paradox* (c. 600 BCE). Epimenides, a Cretan, is reported to have stated that "Cretans are always liars." However, Epimenides' statement is non-paradoxical—in fact false. For if "Cretans are always liars" is true, then Epimenides, a Cretan, has stated a truth, and therefore "Cretans are always liars" must be false. This version of the paradox even appears in the Bible (Titus 1:12–13a): It was one of them, their very own prophet, who said, "Cretans are always liars, vicious brutes, lazy gluttons." That testimony is true.

It seems to have been the Greek philosopher Eubulides of Miletus (fourth century BCE) who first stated the liar paradox in the form in which it has become familiar. Eubulides reportedly posed the question, "A man says that he is lying. Is what he says true or false?"

A version of the paradox occurs in *Don Quixote*, when Sancho Panza is stopped at a bridge and told by the guards that he can pass if he tells the truth and that he will be hanged if he lies; he says, "I shall be hanged." This leaves the guards in a dilemma. For if they let him pass, then he will have lied and so he should have been hanged; while if he is hanged then he will have told the truth and so he should have been allowed to pass.

The liar paradox can also be cast in forms involving more than one statement. For example, consider the following pair of sentences:

(1) the following statement is true
(2) the preceding statement is false.

Assume (1) is true. Then (2) is true. This would mean that (1) is false. Therefore (1) is both true and false.

Assume (1) is false. Then (2) is false. This would mean that (1) is true. Thus (1) is both true and false. So in either case, (1) is both true and false.

There have been a number of attempts to resolve the liar paradox. Here are a few of them.

The great logician Alfred Tarski claimed that liar-type paradoxes could arise only in languages he termed *semantically closed*. By this he meant languages in which it is possible for one sentence to ascribe truth (or falsehood) of another sentence in the same language (or even of itself). To avoid contradiction when discussing truth values of sentences it is necessary, according to Tarski, to divide languages into strict *semantic levels*. When one sentence (S1) refers to another (S2), then S2 must be on a lower semantic level than S1. This prevents sentences from being self-referential since a self-referential sentence would have to be semantically lower than itself. Accordingly, the liar sentence, being self-referential, becomes meaningless in the sense that it cannot be legitimately formulated in any language.

The philosopher Arthur Prior dealt with the liar paradox by denying that there is anything paradoxical about the liar sentence. His analysis turns on the claim that each statement implicitly asserts its own truth. Thus, for example, the statement "it is true that two plus two equals four," contains no more information than the statement "two plus two equals four," because the phrase "it is true that ..." is implicitly present in any statement.

Under this analysis the liar sentence *this sentence is false* now becomes *it is true that this sentence is false*, or *this sentence is true and this sentence is false*. This last sentence is a straightforward contradiction of the form *A and not A*, and hence is false. No paradox now arises because the claim "*A and not A*" *is false*, far from being a contradiction, is actually true.

The philosopher *Graham Priest* and others have suggested that the liar sentence should be considered to be *both true and false*. This is justified by a doctrine known as *dialetheism*, whose principal tenet is the claim that there can be *true contradictions*. Dialethism faces an immediate difficulty arising from the generally accepted law of logic *ex falso quodlibet*, "from a falsehood, anything follows," or *ex contradictione quodlibet*, "from a contradiction, anything follows." According to this law (also known as the *principle of explosion*) any proposition can be deduced from a contradiction. By this law, all propositions follow from a contradiction dialethism accepts, so it would therefore follow that in dialethism *all* propositions would have to be true. In order to avoid this absurd conclusion dialetheists nearly always reject the explosion principle. Logical systems rejecting it are called *paraconsistent*.

In 1931 Gödel used a modified form of the liar paradox to prove his celebrated *Incompleteness Theorem*. This asserts that, in any sound and sufficiently rich mathematical theory *T*—arithmetic or set theory, for example—propositions can always be formulated which one can see to be true but whose truth cannot be proved within the theory. Gödel obtained such a proposition by replacing, in the liar sentence *this sentence is false*, the word "false" by the word "unprovable," producing the sentence *this sentence is unprovable*. This, the Gödel sentence *G*, thus asserts its own *unprovability*. It follows, assuming that *T* is sound, that *G is true but unprovable*. For suppose that *G* were false. Then, since *G* asserts its own unprovability, it would follow that the unprovability of *G* is false, i.e., *G* is provable. But then *G* would be both provable and false, contradicting the soundness of *T*. It follows that *G* must be true. Therefore, since *G* asserts its own unprovability, it must also be unprovable.

Gödel later strengthened the Incompleteness Theorem to show that, for any consistent and sufficiently rich mathematical theory *T*, the consistency of *T* cannot be proved in *T*. If we take *T* to be arithmetic, it follows that, if arithmetic is consistent, then the consistency of arithmetic cannot be proved in arithmetic itself. The great French mathematician André Weil expressed this in the saying *God exists, since mathematics is consistent, and the Devil exists, since we cannot prove it.*

THE LIAR, THE TRUTH-TELLER, AND THE DICE MAN

Related to the liar paradox is the following well-known puzzle. A person travelling to a certain town comes to a fork in the road and doesn't know which branch to take. Two brothers live next to the fork, one of whom always tells the truth and the other whom always lies. What single question should the traveller ask either one of the brothers in order to determine which branch to take? A suitable question is: *which branch would your brother tell me to take?* On hearing the answer to this question, the traveller then does the opposite: if the answer is *the right branch*, he takes the left branch, and if the answer is *the left branch*, he takes the right branch. This works because, if the traveller happens to put the question to the truth-teller, the latter will truthfully report what his lying brother's answer would be, i.e., a lie. Similarly, if the traveller happens to put the question to the liar, the latter will lie about the truthful answer his brother would give, again producing a lie.

A nice way of presenting this is to think of the truth teller (T) as a lens consisting of a simple piece of glass and the liar (L) as a lens in which images appear inverted. Asking the truth-teller what the liar would say is then analogous to looking through the combination of lenses in the order TL. Asking the liar what the truth-teller would say is analogous to looking through the lens combination LT. In both cases the image is inverted, just as, in the original scenario, the answer is always a lie.

Now let's introduce into the scenario a third brother—call him the *dice man*—who *answers entirely at random*. Surprisingly, the traveller can extract the required information by asking just *two* questions. Let us call the truth teller and the liar *determinate*. The first question has then to be phrased in such a way that, *no matter to whom it is addressed*, the answer will enable the traveller to identify a *determinate* brother. Let us say that the truth teller (**T**) is *more truthful* than the dice man (**D**), who is in turn is *more truthful* than the liar (**L**). The traveller picks any of the brothers, *A* say, and labels the other two brothers *B* and *C*. He then asks *A*, *Is B more truthful than C?* If the answer is *yes*, then *C* is determinate. If the answer is *no*, then *B* is determinate. That this is the case can be seen from the following table.

A	B	C	Answer	Determinate brother
T	L	D	No	*B*
T	D	L	Yes	*C*
L	T	D	No	*B*
L	D	T	Yes	*C*
D	L	T	No	*B*
D	L	T	Yes	*C*
D	T	L	No	*B*
D	T	L	Yes	*C*

The essential point here is that the strategy works when *A* is determinate, and *it continues to work when A is the dice man*. This is the case because when *A* is the dice man, both *B* and *C* are determinate, so it actually doesn't matter which one is picked.

Once the traveller has identified a determinate brother, his second question, addressed to the latter, is *which branch would your* determinate *brother tell me to take*? On hearing the answer to this question, the traveller proceeds as in the two-brother case, that is, he does the opposite of what he is told. Thus he achieves his goal.

CURRY'S PARADOX

Like the liar paradox, Curry's paradox (due to the American logician H.B. Curry) involves self-reference.

Consider the sentence, call it *S*:

If the only sentence in this box is true, then God exists.

Is S true? A standard, obviously correct, way to prove the truth of a true conditional sentence is to assume the truth of the antecedent (the "if ..." part) and derive the consequent (the "then ..." part). That is called *conditional proof*. Let's do that.

Assume the antecedent:

(1) The sentence in the box is true.

So we can assert the sentence in the box, which is this:

(2) If the sentence in the box is true, God exists.

Now, from (1) and (2), this clearly follows:

(3) God exists.

But this constitutes a conditional proof of the statement with (1) as antecedent and (3) as consequent:

(4) If the sentence in the box is true, then God exists.

So we've proven that (4) is true.

Now, (4) is the sentence in the box. So we've also proven:

(5) The sentence in the box is true.

And from (4) and (5), this clearly follows:

(6) God exists.

Thus we seem to have proved that God exists. Now clearly if we replace the statement *God exists* by *any* given statement *A*, the same argument will prove that *A* is true. Accordingly we have shown that all statements are true. This absurd conclusion is Curry's paradox.

THE GRELLING-NELSON PARADOX

The *Grelling-Nelson paradox* was formulated in 1908 by the philosophers *Kurt Grelling* and *Leonard Nelson*. It is closely related to Russell's paradox, but it is, perhaps, even more unsettling, since it appears to strike at the very foundations of language.

Define the adjectives *autological* and *heterological* by stipulating that an (English) adjective is *autological* if and only if it is applies to itself and *heterological* if and only if it does not apply to itself. More exactly, an adjective *A* is

- autological if and only if the word "*A*" has the property expressed by the adjective *A*,
- heterological if and only if the word "*A*" does *not* have the property expressed by the adjective *A*.

Clearly an adjective cannot be both autological and heterological.

For instance, the adjectives *polysyllabic* and *English* are autological since the word "polysyllabic" is polysyllabic and the word "English" is English. On the other hand the adjectives *palindromic* and *French* are heterological since the word "palindromic" is not palindromic and the word "French" is not French.

The paradox arises upon asking whether the adjective *heterological* is heterological.

The answer must be yes ("Heterological" is heterological) or no ("Heterological" is not heterological—i.e., it's autological).

(1) Maybe YES: "Heterological" is heterological. Any adjective that's heterological is an adjective that does not apply to itself. So if YES, then "heterological" does not apply to that adjective. So it follows that "heterological" is NOT heterological, and that it's autological. So the YES answer can't be right.

(2) So how about NO: "Heterological" is autological. Any adjective that's autological is an adjective that applies to itself. So if NO, then "heterological" applies to itself. So "heterological" is heterological. So "heterological" is NOT autological. So the NO answer can't be right either.

We reason as follows:

The word "heterological" is heterological if and only if the word "heterological" is heterological,

But since any word that's heterological doesn't apply to itself, i.e., is autological,

The word "heterological" is heterological if and only if the word "heterological" is autological.

This contradiction is the Grelling-Nelson paradox.

As in the case of Russell's bibliography paradox, a deeper analysis dissolves the Grelling-Nelson paradox. To do this, we assume that we are given two collections: a collection of words called *adjectives* and a collection of *properties* of adjectives. If the adjective *a* has the property *P*, we say that *P applies to a*. We also suppose that each property *P* of adjectives is correlated with a unique adjective *a* called its *name*: in that case we say that *a names P*. Now consider the "heterological" property *H* of adjectives which is defined to apply to an adjective *a* just when *a* names a property which doesn't apply to *a*. We now show that *H applies to its own name*. To prove this we argue by contradiction. Write *h* for the name of *H* and suppose that *H* doesn't apply to *h*. Then *h* names a property, i.e., *H*, which doesn't apply to *h*. It now follows from the very definition of *H*, that *H* applies to *h*. We conclude that *H* does indeed apply to *h*. That being established, it follows from the definition of *H* that *h* must also name a property, call it *I*, *which doesn't apply to h*. Then the properties *H* and *I* have the same name (that is, *h*) but they cannot themselves be the same, since *H* applies to its own name but *I* doesn't. We conclude therefore that *there must exist two different properties of adjectives with the same name*. One of these, as we have seen, is the "heterological" property *H*. This is a perfectly well-defined property which doesn't give rise to paradox. *But it must necessarily have the same name as a different adjective.*

Now consider the "autological" property *A* of adjectives which is defined to apply to an adjective *a* just when *a* names a property which *does* apply to *a*. As in the bibliographical case, nothing can be inferred about *A*: it may be autological or heterological; and it may or may not have the same name as a different adjective.

BERRY'S PARADOX

Berry's paradox (apparently due to the librarian G.G. Berry) is an instance of a *paradox of definition*. Consider English expressions defining individual positive integers: for instance *the only even prime number*, which defines the number 2; *the first odd prime number*, which uniquely defines the number 3; *the first number expressible as the sum of two cubes in two different ways*, which uniquely defines the number 1729. For a given number *N* of words, there are finitely many English expressions containing fewer than *N* words, and hence finitely many positive integers definable by expressions containing fewer than *N* words. Since there are infinitely many positive integers, it follows that there must exist positive integers not definable by expressions containing fewer than *N* words, and so there is a *smallest* such integer.

Now consider the integer *n* defined by the expression

> the smallest positive integer not definable by an expression containing fewer than fifteen words.

This expression is fourteen words long, so the integer *n* is defined by an expression containing fewer than fifteen words. This means that *n* *is* definable by an expression containing fewer than fifteen words, and so is *not* the smallest positive integer not definable in under fifteen words. This is a paradox: there must be a positive integer defined by this expression, but since any positive integer it defines is definable in fewer than fifteen words, there cannot be any integer defined by it.

One way of circumventing Berry's paradox is to follow Bertrand Russell and introduce a stratification of defining expressions, into *levels*. The lowest level (0) of defining expressions consists of expressions such as *the third odd number* which contain no reference to defining expressions. At level 1 we will have defining expressions which may involve defining expressions of level 0, such as *the least positive integer greater than three not definable by an expression of level 0 containing less than 27 words*. Level 2 consists of defining expressions which may involve expressions of levels 0, and level 1. In general, level *n* consists of defining expressions involving defining expressions of levels less than *n*. Thus an expression at a given level can involve only expressions of *lower* levels. Distinguishing levels in this way allows paradoxes such as Berry's to be dealt with. There the phrase *definable by an expression* is incorrectly taken to cover not only defining expressions in the usual sense, that is, those of level 0, but also defining expressions of all levels. We must instead insist on specifying the levels figuring in these defining expressions. Thus, in place of the now illegitimate

> the smallest positive integer not definable by an expression containing fewer than fifteen words

we consider

> (*) the smallest positive integer not definable by an expression of level n containing fewer than eighteen words.

This is a defining expression of level $n + 1$, containing fewer than eighteen words. The integer *m* defined by (*) is assuredly *not* definable by an expression *of level n* containing fewer than eighteen words. But *m is* definable by an expression—(*) itself—of level $n + 1$, containing fewer than eighteen words. The contradiction has evaporated.

RICHARD'S PARADOX

The original statement of the paradox, due to the French mathematician Jules Richard (1905), is a paradox of definition concerning real numbers. Here we present a version based just on natural numbers.

Consider a language (such as English) in which the arithmetical properties of positive integers can be defined. For example, *the second positive integer* defines the property of being the natural number two; and *not divisible by any natural number other than 1 and itself* defines the property of being a prime number. While the list of all such possible definitions is itself infinite, it is easily seen that each individual definition is composed of a finite number of words, and hence also a finite number of characters. This being the case, we can order the definitions, first by number of characters and then lexicographically, i.e., in dictionary order. Write **L** for the ordered list of definitions so obtained.

Now we assign a positive integer to each definition in **L** as follows: the definition with the smallest number of characters and alphabetical order is assigned the number 1, the next definition in the sequence is assigned the number 2, and so on. Each definition is assigned a unique integer. Moreover, for each integer *n* there is a unique definition in **L** to which the integer *n* has been so assigned; call that definition the definition *correlated* with *n*. Now it may occasionally happen that an integer actually *possesses* the property characterized by its correlated definition. Suppose, for example, that the number 37 happens to be associated with the definition *not divisible by any integer other than 1 and itself*. Since 37 is not divisible by any integer other than 1 and itself, 37 possesses the property characterized by its correlated definition. Intuitively, however, it would seem much more likely that a given integer fails to have the property characterized by its correlated definition. For example, if the definition *least number that is the sum of two squares* happened to be correlated with the number 28, then 28 does *not* possess the property characterized by its correlated definition. Let us call integers *normal*. Thus a number is normal if and only if it does not possess the property characterized by its correlated definition.

Richard's paradox arises if we take the property of normality to have a definition in **L**. For if this is the case, then that definition—call it *D*—has been assigned an integer *n*, so that *D* is the definition correlated with *n*. A contradiction then immediately arises: for

n has the property characterized by *D*
if and only if *n* is normal
if and only if *n* does not have the property correlated with *n*
if and only if *n* does not have the property characterized by *D*.

The only way of circumventing this paradox would seem to be to conclude that the property of normality does not have a definition in **L**, that is, cannot be finitely definable, that is, defined by means of finitely many words. Now at first sight it appears that normality *has* been finitely defined. But has it? The definition of normality implicitly depends on isolating those expressions in English which define natural numbers. Thus, for normality to be finitely definable, it is necessary that *the property of being an expression defining a positive integer*—call it *P*—be finitely definable.

Now we can certainly correlate expressions with integers by following the same procedure as we did for definitions. The property of being an expression defining a positive integer then corresponds to a property, call it *Q*—of the so correlated integers. Thus an integer *n* has the property *Q*, that is, *n* is an integer corresponding to an expression that defines a positive integer, exactly when the expression correlated with *n* has the property *P*, that is, is itself an expression that defines a positive integer. Clearly, then, the property *P* of expressions is finitely definable if and only if the property *Q* of integers is finitely definable. But there is no reason to suppose that *Q* is itself finitely definable. And in fact Richard's paradox shows that *Q* cannot be finitely definable. To put the conclusion succinctly, *the property of being finitely definable is not itself finitely definable.*

THE PARADOX OF THE HEAP

The *paradox of the heap* or *sorites paradox* (from the Greek *sōritēs* "heap")—attributed to the ancient Greek philosopher Eubulides of Miletus—arises from the vagueness of certain predicates in ordinary language. In a typical formulation, we consider a heap of sand, from which grains are removed one by one. The paradox arises when one considers what happens when the process is repeated sufficiently many times. For suppose we make the natural assumption that, if we remove a single grain from a heap, we are still left with a heap. Then eventually just a single grain remains: is it still a heap? Or are even no grains at all a heap? If not, when did the heap change into to a non-heap?

We can turn the paradox on its head by starting with a totally bald man, and, noting that any man with just one more hair than a bald man is still bald, conclude that every man must be bald. For a man with no hair is bald, so a man with just one hair is bald, and thus a man with two hairs is bald, ... whence a man with any number of hairs is bald.

A related formulation of the paradox is to suppose given a set of coloured chips such that the variation in colour of two adjacent chips is too small—a difference in wavelength of 1 nanometre say—for the human eye to be able to distinguish between them. Suppose that the first chip is coloured violet, which has a wavelength of about 400 nanometres, and the last chip is coloured red, with a wavelength of 650 nanometres. If we assume, as in the case of the bald man, that a chip whose colour differs in wavelength by one nanometre from a violet coloured chip would still be seen as violet, then the 'bald man' argument leads to the conclusion that the red chip would also have to be seen as violet.

The paradox can be reconstructed for a variety of predicates—all of which can be seen to be vague—for example, with "short," "poor," "young," "red," and so on.

A natural response to the paradox is to introduce a "fixed boundary" to the concept of heap by defining a "heap" to be a set of grains containing at least a certain fixed number—10000, say—of grains. In that case, a set of 9999 grains is not a heap but one of 10000 is. This seems unnatural since there would appear to be little significance to the difference between 9999 grains and 10000 grains. Wherever the boundary is set, it remains arbitrary. A more acceptable, if radical solution would be to call *any* collection of two or more a heap!

The paradox can be given a striking formulation by generalizing the colour example above. Suppose that we have a set S of things—colours, collections of grains of sand, people—on which is defined a relation **I** which we shall call *indistinguishability*. So for two elements a and b of S, a and b will be in the relation **I**—which we write as $a\mathbf{I}b$—if and only if a is indistinguishable from b. We shall suppose that **I** is *reflexive*—for any a, $a\mathbf{I}a$—and *symmetric*—for any a, b, $a\mathbf{I}b$ if and only if $b\mathbf{I}a$. Let us call a property (or predicate) P defined on S *vague* if it is preserved under indistinguishability, that is, if $a\mathbf{I}b$ and $P(a)$ then $P(b)$ (in words: *anything indistinguishable from something with the property P also has the property P*). Let us say that two elements a, b of S are *connected* if there is a sequence a_0, ..., a_n of elements of S such that $a_0 = a$, $a_n = b$ and, for each i, a_i,$\mathbf{I}a_{ii+1}$. Call S *connected* if each pair of elements of S are connected.

Suppose now that S is connected. Then, *for any vague property P on S, if* some *element of* S *has P, then* every *element of S has P*. To see this, suppose that a is an element of S such that a has the property P—we write this as $P(a)$—and let b be an arbitrary element of S. Then since S is connected, there is a sequence $a_0, ..., a_n$ of elements of S such that $a_0 = a$, $a_n = b$ and, for each i, $a_i,\mathbf{I}a_{ii+1}$. Now since $P(a)$, i.e., $P(a_0)$ and $a_0\mathbf{I}a_{i1}$, it follows from the vagueness of P that $P(a_1)$. From this it follows similarly that $P(a_2)$, whence $P(a_3)$ and so on. Finally we obtain $P(a_n)$, i.e., $P(b)$. Since b was arbitrary, we conclude that every element of S has P.

From this we infer that a *vague property either applies to everything, or it applies to nothing.* For example, consider the case of the vague predicate "bald" or better, *baldish*. Here S is the set of (heads of) men and **I** is the relation of differing by at most one hair. Then, if there is at least one baldish man, all men are baldish—including Brad Pitt. If, on the other hand, there is at least one nonbaldish man, then all men are nonbaldish—including Bruce Willis.

Appendix 2

Reflections on the Constant and the Changing

The world, as we perceive it, is in a perpetual state of flux. It has been justly remarked that change is the only constant, that everything is subject to change apart from change itself. The understanding of change has been a problem for philosophers since antiquity.

The opposition of the Constant and the Changing is closely connected with that of the One and the Many. A thing can change through time, so appearing to become many different things, and yet in some sense remain that same thing. For example, a human being begins as an infant, becomes a child, grows into an adolescent, and eventually becomes an adult. Infant, child, adolescent, adult—all are physically different, and yet we accept that they are all the *same person*. The underlying identity of the person—the Constant—persists through the phases of his or her development—the Changing.

Monists agree that all Being is One, but they can differ markedly on the issue of the Constant and the Changing. Parmenides and Heraclitus were both monists, but while Parmenides denied absolutely the possibility of change, Heraclitus maintained that all things are in a state of flux—famously declaring that *no man ever stepped in the same river twice*.

From the early seventeenth century mathematicians have striven to provide the phenomenon of change with an appropriate mathematical formulation.

Consider a familiar form of change: *change of location*, or *motion*. The fundamental nature of this form of change led the atomist philosophers of ancient Greece and their successors in the eighteenth and nineteenth centuries to claim that all forms of change in Nature were reducible to the motion of material bodies. Now motion, considered as a form of change, is itself reducible to a still more fundamental form of change—*spatiotemporal* change. When a billiard ball is struck, the billiard ball itself does not change, but its location in space and the time at which it occupies that location is subject to continual change. It was not in fact until the seventeenth century that motion came to be conceived as a dependence of variable spatial position on variable time, and so as a *functional relation* between space and time. Lacking an adequate formulation of this idea, the mathematicians of Greek antiquity were unable to produce a satisfactory mathematical analysis of motion, or more general forms of change. The problem of analyzing motion was also compounded by Zeno's paradoxes, which, as we have remarked, were intended to show that motion was impossible.

The incorporation of variation into mathematics in the seventeenth century led to the triumphs of the calculus, mathematical physics, and the mathematization of our understanding of nature. But difficulties surfaced in the attempt to define the *instantaneous rate of change* of a varying quantity—the fundamental concept of the differential calculus. Like the ancient Pythagorean effort to reduce the continuous to the discrete, the endeavour of the mathematicians of the seventeenth century to reduce the varying to the static through the use of infinitesimals led to outright *contradictions*.

The response of their successors was, effectively, to replace change by constancy. Cantor in particular abandoned the concept of a varying quantity in favour of a completed, static *domain of variation*—the *linear continuum*—itself to be regarded as a set of atomic individuals—thus, like the Pythagoreans, at the same time replacing the continuous by the discrete. He also banished infinitesimals and the idea of geometric objects as being generated by points or lines in motion.

Philosophers such as Franz Brentano and Charles S. Peirce in the late nineteenth century, and mathematicians such as L.E.J. Brouwer and Hermann Weyl, in the early twentieth century, raised spirited objections to the idea of treating the continuum as a collection of discrete points. They were convinced that the truly continuous was not reducible to the discrete.

It was with Brouwer that logic itself enters the fray. Rejecting the Cantorian account of the continuum as discrete, Brouwer identified points on the linear continuum—or *real numbers*—as entities "in the process of becoming," that is, as embodying a certain kind of change. Brouwer saw that the variable nature of these entities would cause them to be at variance with certain laws of logic which had been affirmed since Aristotle. The most important of these is the *law of excluded middle* or *bivalence*—"each statement is either true or false."

To get an idea of how the law of excluded middle can fail in Brouwer's conception of the real numbers, consider the assertion, call it *A*: *the sequence 123456789 occurs in the decimal expansion* of π. We can calculate π to as many decimal places as we please, and if in *actuality* the sequence 123456789 occurs *somewhere*, i.e., *A* is true, then we shall discover this fact after a sufficient amount of calculation. But if the sequence occurs *nowhere*, that is, *A* is false, no amount of calculation can ever establish this beyond doubt. For Brouwer this meant that we cannot be *certain* that *A* is either true or false, since we cannot decide *in advance* which alternative holds. The failure of the law of excluded middle arises only in the case of mathematical entities, such as the decimal expansion of π, which are always "in the process of becoming," potentially infinite objects whose formation goes on forever and never stops. Suppose that in place of the assertion *A* we consider the assertion *B*: *the sequence 123456789 occurs before the* $10^{10^{10}}$*th place in the decimal expansion* of π. Then in order to be certain of whether *B* is true or false we need only(!) calculate π to $10^{10^{10}}$ places. This is a *finite* procedure and could, in principle be carried out. In that case, according to Brouwer, we can be *certain* that *B is* either true or false. This is the case even if we have not *actually* determined which alternative holds. The point is that *in principle* we could determine which one holds.

This revolutionary insight led to a new form of logic, *intuitionistic logic*—a "non-Aristotelian" logic in which the law of excluded middle is no longer affirmed. It was later shown that the intuitionistic logic of

Brouwer is compatible with a very general concept of change, embracing all forms of continuous variation, and which is in particular compatible with the use of infinitesimals.

Certain philosophers—notably the nineteenth-century thinkers G.W.F. Hegel and Karl Marx—believed that a true understanding of the phenomenon of change would require the development of a dialectical logic or "logic of contradiction," in which the *law of noncontradiction*—"no statement can be both true and false"—is repudiated. It is a striking fact that, so far at least, more light has been shed on the problem of change by challenging the law of excluded middle rather than by questioning the law of non-contradiction.

Appendix 3

Oppositions in Kant's Philosophy

A number of oppositions (Finite/Infinite, One/Many, Continuous/Discrete, and, arguably, Constant/Changing) play significant roles in the philosophy of Immanuel Kant (1724–1804). Kant's mature philosophy, known as *transcendental idealism*, rests on the division of reality into two realms. The first, the *phenomenal* realm, consists of appearances or objects of possible experience. The second, the *noumenal* realm, consists of "entities of the understanding to which no objects of experience can ever correspond." These are known as *things-in-themselves*. According to Kant, a thing-in-itself, signifies "only the thought of something in general, in which I abstract from everything that belongs to the form of sensible intuition." A thing-in-itself is thus a pure entity from which all qualities have been abstracted.

Objects of the phenomenal realm, the appearances, occupy space and time and so are continuous and limitlessly divisible. Kant seems to have regarded things-in-themselves as discrete entities in the sense of not being limitlessly divisible, and hence as being compounded from atoms. This may be inferred from his assertion in the *Metaphysical*

Foundations of Natural Science of 1786 that a thing-in-itself "must in advance already contain within itself all the parts in their entirety into which it can be divided." So were a thing-in-itself to be limitlessly divisible, one could infer that it consists of an infinite multitude of parts. This, however, is impossible "because there is a contradiction involved in thinking of an infinite number as complete, inasmuch as the concept of an infinite number already implies that it can never be wholly complete." Here we see an instance of the Finite/Infinite opposition.

Kant's *Critique of Pure Reason* (1781) is regarded as one of the most important works in modern philosophy. Central to the work are the so-called *Antinomies*. These are contradictions which Kant held to be unavoidable consequences of any attempt to grasp the nature of reality.

In the *Critique* Kant brings a new subtlety to the analysis of the opposition between the Continuous and the Discrete. This may be seen in the second Antinomy, which concerns the question of the composition of matter, or extended substance. Is it (*a*) discrete, that is, does it consist of simple or indivisible parts, or (*b*) continuous, that is, does it contain parts within parts *ad infinitum*? Although (*a*), which Kant calls the *Thesis* and (*b*) the *Antithesis* would seem to contradict one another, Kant offers proofs of both assertions. The resulting contradiction may be resolved, he asserts, by observing that while the Antinomy "relates to the division of appearances," the arguments for (*a*) and (*b*) implicitly treat matter or substance as things-in-themselves:

> For these [i.e., appearances] are mere representations; and the parts exist merely in their representation, consequently in the division (i.e., in a possible experience where they are given) and the division reaches only so far as such experience reaches. To assume that an appearance, e.g., that of a body, contains in itself before all experience all the parts which any experience can ever reach is to impute to a mere appearance, which can exist only in experience, an existence previous to experience. In other words, it would mean that mere representations exist before they can be found in our faculty of our representations. Such an assumption is self-contradictory, as also every solution of our misunderstood problem, whether we maintain that bodies in themselves consist of an infinite number of parts or of a finite number of simple parts.

Kant concludes that both Thesis and Antithesis "presuppose an inadmissible condition" and accordingly "both fall to the ground, inasmuch as the condition, under which alone either of them can be maintained, itself falls."

Kant identifies the inadmissible condition as the implicit taking of matter as a thing-in-itself, which in turn leads to the mistake of taking the division of matter into parts to subsist independently of the act of dividing. In that case, the Thesis implies that the sequence of divisions is finite; the Antithesis, that it is infinite. Now if matter or substance is a thing-in-itself, and not just an appearance, then the *completed* (or at least completable) sequence of divisions would have an objective existence. But in that case the sequence of divisions would have to be—*in actuality*—both finite and infinite. This is a contradiction, and it follows that matter and extended substance cannot be things-in-themselves; that is, matter and substance are mere appearances. And for appearances, Kant maintains, divisions into parts are not completable in experience, with the result that such divisions can be considered *neither finite nor infinite*:

> We must therefore say that the number of parts in a given appearance is in itself neither finite nor infinite. For an appearance is not something existing in itself, and its parts given in and through the regress of the decomposing synthesis, a regress which is never given in its absolute completeness, either as finite or infinite.

It follows that, for appearances, both Thesis and Antithesis are *false*.

Later in the *Critique* Kant argues that, while each part generated by a sequence of divisions of an intuited whole is given with the whole, the sequence's incompletability prevents *it* from forming a whole, from which it follows that no such sequence can be claimed to be actually infinite. This is an instance of both the Finite/Infinite and the Whole/Part oppositions.

The better-known First Antinomy in the *Critique* concerns space and time. It can be seen as an embodiment of both the Finite/Infinite and the Bounded/Unbounded oppositions. It is presented in the form of a Thesis:

> the world has a beginning in time, and is also limited as involves space;

and an Antithesis:

> the world has no beginning in time, and no limits in space—it is infinite as regards both time and space.

As far as the temporal aspect of the First Antinomy is concerned, Kant conceives of the world's past as a series, and he presents the following argument for the finiteness of that past:

> The infinity of a series consists in the fact that it can never be completed through successive synthesis. It thus follows that it is impossible for an infinite world-series to have passed away.

By "successive synthesis" we may understand "temporal succession." In that case, what does it mean to say that the series *P* of past events is "completed" through temporal succession? A possible meaning could be that there is a class *E* of past events such that every past event is obtained by temporal succession from the events in *E*. Even granting this, however, one can infer that *P* is finite only if one assumes that *E is also finite*. In fact these two assertions are actually *equivalent*, and, assuming either of them, one can take *E* to consist of a single "initial" event which may be identified with the origin of time. But it is obviously consistent for *E*, and hence also *P*, to be infinite, and so Kant's argument, construed in this way, fails.

The situation boils down, it seems, to this. If the past *were* actually finite, it is of course generated by temporal succession from an initial event. But if it is infinite, although it *grows* by temporal succession, it is not generated or completed by temporal succession from any finite set of events. In this case the infinite past must be conceived at any instant as being *given as a whole*, and only *finite* parts of it as being "completed" by temporal succession.

Jonathan Bennett, in his thought-provoking paper, *The Age and Size of the World*,[1] sums up the conclusion of the First Antinomy as follows:

> Although Kant denies that the world can be infinitely old or large, he thinks that it cannot be finitely large or old either.

1 *Synthese* 23 (1971): 127–46.

In explicating this assertion, Bennett concludes that what Kant means by "the world is not finite in size" is "no finite amount of world includes all the world there is," or "every finite quantity of world excludes some world." Bennett submits that this last statement "seems to Kant to be a weaker statement than the statement that there is an infinite amount of world." More generally, Bennett suggests that

> Kant is one of those who think that
>
> Every finite set of F's excludes at least one F, (1)
>
> though it contradicts the statement that there are only finitely many F's, is nevertheless weaker than
>
> There is an infinite number of F's (2).

Bennett implies that Kant is simply mistaken here, that in fact (1) and (2) are equivalent.

But is this right? Let us bring to bear some contemporary mathematical ideas on the matter.

Call a set A *finite* if for some natural number n, all the members of A can be enumerated as a list $a_0, \ldots, a_n$; *potentially infinite* if it is not finite, that is, if, for any natural number n, and any list of n members of A, there is always a member of A outside the list; *actually infinite* if there is a list $a_0, \ldots, a_n, \ldots$ (one for each natural number n) of *distinct* members of A; and *Kantian* if it is potentially, but not actually infinite, that is, if it is neither finite nor actually infinite.[2] Now *is it possible for a set to be Kantian*, just as Kant (according to Bennett) thought the actual world was?

For suppose that we are given a potentially infinite set A, and we attempt to show that it is actually infinite by arguing as follows. We start by choosing a member a_0 of A; since A is potentially infinite, there must be a member of A different from a_0; choose such a member and call it a_1. Now again by the fact that A is potentially infinite, there is a member of A different from a_0, a_1—choose such and call it a_2. In

2 In contemporary set-theoretic terminology, my term "potentially infinite" corresponds to "infinite"; "actually infinite" to "transfinite" or "Dedekind infinite"; and "Kantian" to "infinite Dedekind finite."

this way we generate a list $a_0, a_1, a_2, \ldots$ of distinct members of A; so, we are tempted to conclude, A is actually infinite. But clearly the cogency of this argument hinges on our presumed ability to "choose," for each n, an element of A distinct from $a_0, \ldots, a_n$. This ability is underpinned by the axiom of choice, the set-theoretical principle we discussed in Chapter II. As we have pointed out, the axiom of choice was shown by Gödel in 1938 to be consistent with the principles on which set theory rests. Nevertheless, in 1964 Paul Cohen showed that its *denial* is equally consistent with those principles. In fact, the axiom of choice can be denied in such a way as to prevent the argument just presented from going through, that is, to allow the presence of potentially infinite sets which are not at the same time actually infinite. That is, the existence of Kantian sets is consistent with the axioms of set theory (and classical logic) *as long as the axiom of choice is not assumed.*

The axiom of choice may be regarded as a principle ensuring that the domain of sets is "static" in the sense that families of sets "sit still" long enough to enable elements to be "chosen" from them. Accordingly the existence of "Kantian" sets is compatible with set theory, as long as the domain of sets has not been rendered "too static" through the imposition of the axiom of choice.

Another, more direct way of obtaining a Kantian set is to allow "sets" to undergo *explicit variation.* Let us consider variation over *discrete time*, as traced by the ticks of a clock marking seconds. The clock's ticks can be represented by the sequence of positive integers 1, 2, ... Suppose now that we assemble a set of matchsticks by starting with a single matchstick in the first second, adding another matchstick in the next second, etc. Then by the nth second we will have assembled a set of n matchsticks. We can think of the set of matchsticks as growing, or *varying*, with each tick of the clock. This is a simple example of a set varying over discrete time. This set—call it **Matchstick**—grows with each successive tick of the clock and is hence potentially infinite. On the other hand at each specific time it is finite and so is not actually infinite. In the realm of "sets varying through discrete time," **Matchstick** is a Kantian set.[3]

3 *Cognoscenti* will recognize that "realm of sets varying over discrete time" is in fact the topos of sets varying over the natural numbers. But within this topos not only the axiom of choice but also the law of excluded middle fails, a course that Kant would likely have found most unpalatable.

It would seem then that Kant's conclusion in the First Antinomy that space and time are neither finite nor infinite—that is, potentially infinite—*is* coherent (or at least consistent) after all. This applies equally to the Second Antinomy, at least in respect of Kant's contention that ongoing sequences of divisions of appearances cannot be assumed to terminate, and are accordingly neither finite nor infinite. But these assertions can only make sense in a conceptual framework conceived as undergoing some form of variation, in which the axiom of choice must necessarily fail.

Appendix 4

The Principle of Microstraightness, Nilpotent Infinitesimals, and the Differential Calculus

It is instructive to see how the two concepts of geometric infinitesimal introduced in Chapter I are related, and how they naturally give rise to the idea of a nilpotent infinitesimal. Let us suppose we are given a smooth curve *AB*, and we want to evaluate the area of the region *ABCO* by regarding as the sum of thin rectangles *XYRS* (exaggerated for clarity in the figure on the following page). If *X* and *Y* are *distinct* points then so are *Y* and *R*, so that the area Δ under the curve is nonzero. That being the case, the figure *ABCO* cannot *literally* be the sum

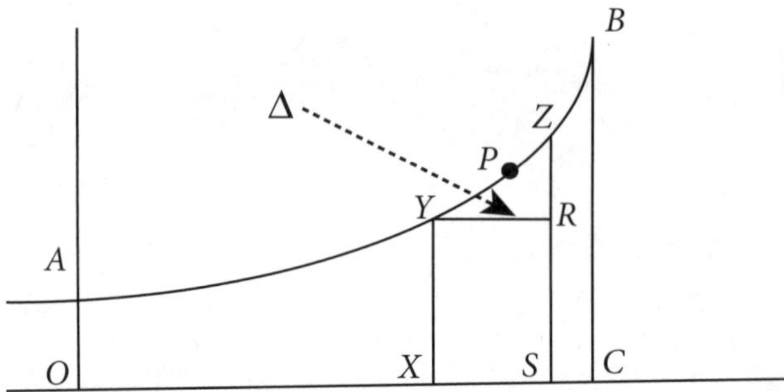

of such rectangles as *XYRS*. On the other hand, if *X* and *S coincide*, then Δ is zero, but *XYRS* contracts into a straight line with no area. In order, therefore, for *ABCO* to be the sum of rectangles like *XYRS*, we must require that their base vertices *X*, *S* be *indistinguishable without coinciding, and yet the area* Δ *be zero*. This necessitates that the segment *XS be a linear infinitesimal of a special kind*: let us follow Barrow and call it a *linelet*.

Let us call a figure *nondegenerate* if it does not coincide with a single point. To achieve our object we want *YRZ* to be a nondegenerate triangle of *zero area*. For this to be the case we must clearly require first that

(a) the segment *YZ* of the curve around the point *P* is actually straight and nondegenerate (in particular does not reduce to *P*).

In this event, the area Δ of *YRZ* is, by elementary geometry, proportional to the square of the length of the line *XS*, so that, if this area is to be zero, we must further require that

(b) *XS* is a nondegenerate line the square of whose length is zero, that is, *the length of the linelet XS is a nilpotent infinitesimal.*

If, for any point *P*, a segment *YZ* of the curve exists such that the conditions (a) and (b) are satisfied, then the rectangles *XYRS* may be regarded as indivisibles whose sum makes up the figure. Thus, an indivisible of the figure may be regarded as a rectangle with a linelet—whose length is a nilpotent infinitesimal—as its base.

Now if this procedure is to apply to any curve, (a) needs to be extended to the following principle:

I. For any smooth curve C and any point P on it, there is an infinitesimal nondegenerate segment of C—let us call it a *microsegment*—around P which is *straight*.

And (b) must be extended to the following principle

II. The set D of nilpotent infinitesimals is nondegenerate, i.e., does not reduce just to zero.

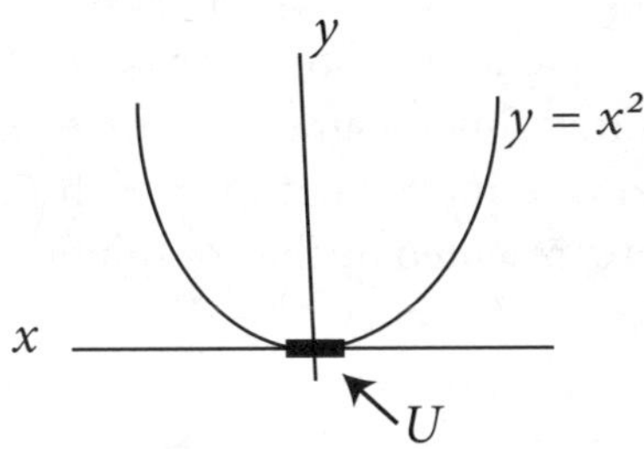

Principle II is actually a consequence of Principle I. For consider the curve above with equation $y = x^2$. Assuming Principle I, let U be the straight segment of the curve around the origin: U is the intersection of the curve with its tangent at the origin (i.e., the x-axis). Thus U is the set of points x satisfying $x^2 = 0$. In other words, U and D are identical. Since Principle I asserts the nondegeneracy of U, i.e., D, we obtain Principle II.

Principle I is called the *Principle of Microstraightness* for smooth curves. It asserts, in essence, that every curve can be considered as being "locally straight." As we have seen, it entails the presence of nilpotent infinitesimals. But it is much more powerful than that.

To begin with, it is closely related both to *Leibniz's Principle of Continuity* and the *Principle of Microuniformity* of natural processes. Leibniz's principle, in essence, is the assertion that processes in nature occur continuously. The Principle of Microuniformity further asserts that any such process can be considered as taking place at a constant rate over an infinitesimal interval of time, i.e., over one of Barrow's "timelets." (Here we see an instance of the opposition between the Constant and the Changing.) For example, if the process is the motion of a body, the Principle of Microuniformity entails that during an infinitesimal interval of time the body experiences no accelerations.

This idea, although rarely explicitly stated, is freely employed informally in classical mechanics and the theory of differential equations.

The close relationship between the Principles of Microuniformity and Microstraightness becomes clear when natural process—for example, the motions of bodies—are represented as curves correlating dependent and independent variables. For then, microuniformity of the process is directly represented by microstraightness of the associated curve.

The Principle of Microstraightness offers an intuitively satisfying account of *motion*. For it entails that infinitesimal segments of (the curve representing) a body's motion are not degenerate points, at which, as in Zeno's paradox of the Arrow, no movement is possible. Rather, these infinitesimal segments—these *microsegments*—are just large enough to make movement over each detectable. On this reckoning, motion is represented by the continuously varying straight microsegment of its associated curve.

A straight microsegment of a curve may be thought of as an infinitesimal "rigid rod," just long enough to point in a direction but too short to bend. It is like an infinitesimal speedometer needle, an entity possessing *direction without length*, intermediate in nature between a point and a straight line.

This "infinitesimal" analysis can also be applied to elucidate the nature of (subjective) *time*. Objectively, time is represented as a succession of discrete instants, isolated "nows" at which time has, as it were, stopped. The Principle of Microstraightness suggests rather that time should be regarded as an assemblage of smoothly overlapping "timelets." Each timelet may be considered to represent a "now" (or what psychologists call a "specious present")—a present moment over which time is still passing.

The purely mathematical power of the Principle of Microstraightness is demonstrated by the elegant way in which it resolves the problem of assigning a quantitative meaning to the concept of *instantaneous rate of change*—as already pointed out, the fundamental goal of mathematicians in creating the differential calculus. For, given a smooth curve C representing a physical process—a motion, say—the instantaneous rate of change of the process at a point P on C is given simply by the *slope*[1] of the straight microsegment ℓ forming part of the curve at P (ℓ is part of the tangent to C at P).

1 The slope of a line is defined to be the tangent the line makes with the horizontal, i.e., the x-axis.

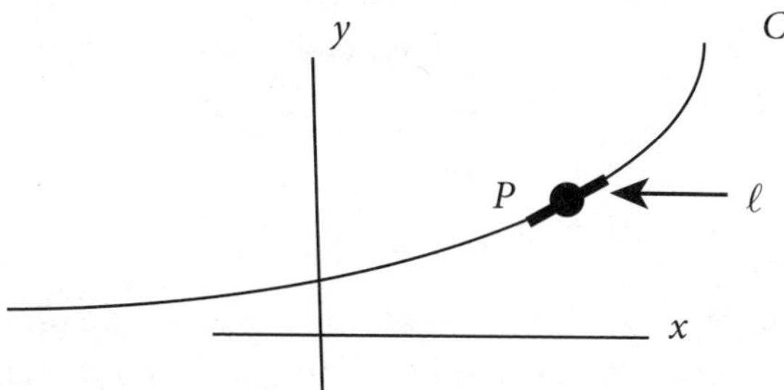

Suppose now that C is given by the equation $y = f(\mathrm{x})$ and P has coordinates (x_0, y_0). Then the slope of the tangent to the curve at P is, according to the usual definition in the differential calculus, the value $f'(x_0)$ of the *derivative* f' of f at x_0. The presence of nilpotent infinitesimals guaranteed by the Principle of Microstraightness enables this value to be calculated in a straightforward manner, as the following argument shows.

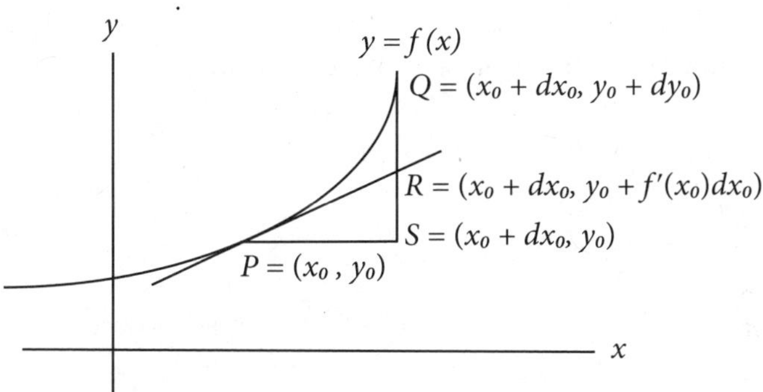

Let dx_0 be any small change in x_0. The corresponding small change dy_0 in $y_0 = f(x_0)$ is represented by the line SQ in the figure immediately above. This may be split into two components. The first component is the change in y_0 along the tangent to the curve at P; this is represented by the line SR, which has length $f'(x_0)dx_0$ (= slope of PR × length of PS). The second component is represented by the line QR; we write its length in the form $\mathrm{H}(dx_0)^2$, where H is some number whose value depends on x_0 and dx_0. The length of SQ is the sum of the lengths of SR and QR, so that

$$f(x_0+dx_0) - f(x_0) = dy_0 = f'(x_0)dx_0+H(dx_0)^2.$$

Now suppose that dx_0 is a nilpotent infinitesimal, which for simplicity we shall denote by the Greek letter ε. Then $\varepsilon^2=0$ and the above equation reduces to

$$f(x_0+\varepsilon)-f(x_0) = f'(x_0)\varepsilon.$$

If we now allow x_0 to vary, this last equation may be understood as asserting that the change in the value of $f(x_0)$ attendant on a nilpotent infinitesimal change ε in x_0 is precisely equal to $f'(x_0)\varepsilon$. The derivative $f'(x_0)$ of f at x_0 thus determined as the *unique* number A satisfying the equation

$$(1)\ \ f(x_0+\varepsilon)-f(x_0) = A\varepsilon$$

for all nilpotent infinitesimals ε. This equation can be used to derive the basic rules of the differential calculus.

First, we use (1) to determine the derivative of the function x^n Write A for the derivative of $f(x) = x^n$ at $x_0 = a$. Then, by (1), A is the unique number such that, for all nilpotent infinitesimal ε,

$$(2)\ \ (a+\varepsilon)^n - a^n = f(a+\varepsilon)-f(a) = A\varepsilon.$$

Now, by the binomial theorem,

$$(a+\varepsilon)^n = a^{n-1} + a^{n-1}\varepsilon + \text{terms in } \varepsilon^2 \text{ and higher powers.}$$

Since $\varepsilon^2 = 0$ the terms in ε^2 and higher powers all vanish, so that

$$(a+\varepsilon)^n = a^{n-1} + na^{n-1}\varepsilon.$$

It now follows from this and (2) that

$$na^{n-1}\varepsilon = (a+\varepsilon)^n - a^n = A\varepsilon.$$

Since A is the unique number satisfying (2) for all ε, it follows that

$$A = na^{n-1}.$$

Accordingly the derivative of x^n at $x = a$ is na^{n-1}.

Finally, we deduce *Leibniz's product rule* for derivatives. Suppose we are given two functions $y = f(x)$ and $y = g(x)$. We want to determine the derivative of the function $h(x)$ given by $h(x) = f(x) \times g(x)$. First, we observe that, if A, B are the derivatives of f, g at $x = a$, then by (1), for arbitrary ε,

$$f(a+\varepsilon) - f(a) = A\varepsilon$$
$$g(a+\varepsilon) - g(a) = B\varepsilon$$

so that

$$(3) \quad \begin{aligned} f(a+\varepsilon) &= f(a) + A\varepsilon \\ g(a+\varepsilon) &= g(a) + B\varepsilon. \end{aligned}$$

Now write C for the derivative of h at $x = a$. Then C is the unique number such that, for all ε,

$$(4) \quad h(a+\varepsilon) - h(a) = (f \cdot g)\,(a+\varepsilon) - f(a) \cdot g(a) = C\varepsilon.$$

Now, using (3)

$$\begin{aligned} (f \cdot g)\,(a+\varepsilon) &= f(a+\varepsilon) \cdot g(a+\varepsilon) \\ &= [f(a) + A\varepsilon] \cdot [g(a) + B\varepsilon] \\ &= f(a) \cdot g(a) + [Ag(a) + Bf(a)]\varepsilon + AB\,\varepsilon^2. \end{aligned}$$

Since $\varepsilon^2 = 0$, we obtain

$$(f \cdot g)\,(a+\varepsilon) - f(a) \cdot g(a) = [Ag(a) + Bf(a)]\varepsilon.$$

But C is the unique number satisfying (4) for all ε.

It follows that

$$C = Ag(a) + Bf(a).$$

Suppressing a, and writing, as is usual in the differential calculus, f' for A, g' for B, and $(f \cdot g)'$ for C, this last equation becomes Leibniz's product rule

$$(f \cdot g)' = f' \cdot g + g' \cdot f.$$

Further Reading

SET THEORY AND THE INFINITE

José Bernardete, *Infinity*. Oxford: Clarendon Press, 1964.
R. Courant and H. Robbins, *What Is Mathematics?* New York: Oxford University Press, 1996.
E. Kasner and J.R. Newman, *Mathematics and the Imagination*. Mineola, NY: Dover Publications, 2001.
Morris Kline, *Mathematics: The Loss of Certainty*. New York: Oxford University Press, 1982.
A.W. Moore, *The Infinite*. London: Routledge, 2001.
Brian Rotman, *Ad Infinitum: The Ghost in Turing's Machine*. Redwood City, CA: Stanford University Press, 1993.
Bertrand Russell, *Introduction to Mathematical Philosophy*. Charleston, SC: Nabu Press, 2010.
Bertrand Russell, *The Principles of Mathematics*. New York: W.W. Norton, 1996.
Barnaby Sheppard, *The Logic of Infinity*. Cambridge: Cambridge University Press, 2014.
N. Ya Vilenkin, *In Search of Infinity*. Boston: Birkhauser, 1995.

THE CONTINUOUS AND THE DISCRETE

John L. Bell, *The Continuous and the Infinitesimal in Mathematics and Philosophy*. Milan: Polimetrica, 2005.

ZENO'S PARADOXES

A. Grunbaum, *Modern Science and Zeno's Paradoxes*. London: George Allen and Unwin, 1968.

INFINITESIMALS

John L. Bell, *A Primer of Infinitesimal Analysis*, 2nd ed. New York: Cambridge University Press, 2008.

J.M. Henle and E.M. Kleinberg, *Infinitesimal Analysis*. Mineola, NY: Dover Publications, 2003.

NON-EUCLIDEAN GEOMETRY

Jeremy Gray, *Ideas of Space: Euclidean, Non-Euclidean, and Relativistic*, 2nd ed. Oxford: Oxford University Press, 1989.

Marvin J. Greenberg, *Euclidean and Non-Euclidean Geometry: Development and History*. New York: W.H. Freeman, 2007.

TIME TRAVEL

J. Richard Gott, *Time Travel in Einstein's Universe: The Physical Possibilities of Travel Through Time*. New York: Mariner Books, 2002.

Susan Schneider (ed.), *Science Fiction and Philosophy: From Time Travel to Superintelligence*. Chichester, UK: Wiley-Blackwell, 2010.

RELATIVITY THEORY

Max Born, *Einstein's Theory of Relativity*. Mineola, NY: Dover Publications, 1962.

Albert Einstein, *Relativity: The Special and the General Theory*. Woodland, CA: Ancient Wisdom Publications, 2010.

N. David Mermin, *It's About Time: Understanding Einstein's Relativity*. Princeton, NJ: Princeton University Press, 2009.

Hans Reichenbach, *The Philosophy of Space and Time*. Mineola, NY: Dover Publications, 1957.

E.F. Taylor and J.A. Wheeler, *Spacetime Physics*. New York: W.H. Freeman, 1966.

Kip S. Thorne, *Black Holes and Time Warps: Einstein's Outrageous Legacy*. New York: W.W. Norton, 1995.

QUANTUM THEORY

Jim Baggott, *The Quantum Story*. Oxford: Oxford University Press, 2013.

William H. Cropper, *The Quantum Physicists and an Introduction to Their Physics*. New York: Oxford University Press, 1970.

W. Heisenberg, *Physics and Philosophy: The Revolution in Modern Science*. New York: Harper Perennial, 2007.

A. Rae, *Quantum Physics: Illusion or Reality?* Cambridge: Cambridge University Press, 1986.

COSMOLOGY

Brian Clegg, *Before the Big Bang: The Prehistory of Our Universe*. New York: St Martin's Press, 2009.

Paul Davies, *Space and Time in the Modern Universe*. Cambridge: Cambridge University Press, 1977.

Paul Davies, *Other Worlds*. London: Abacus, 1982.

Brian Greene, *Parallel Universes and the Deep Laws of the Cosmos*. London: Vintage, 2011.

Martin Rees, *Just Six Numbers: The Deep Forces That Shape the Universe*. New York: Basic Books, 2001.

Alex Vilenkin, *Many Worlds in One: The Search for Other Universes*. New York: Hill and Wang, 2007.

INTUITIONISTIC MATHEMATICS

Michael Dummett, *Elements of Intuitionism*, 2nd edition. Oxford University Press, 2000.

A. Heyting, *Intuitionism: An Introduction*. Amsterdam: North-Holland Publishing Co., 1971.

PARADOXES IN GENERAL

R.M. Sainsbury, *Paradoxes*, 3rd ed. Cambridge: Cambridge University Press, 2009.
Roy Sorensen, *A Brief History of the Paradox: Philosophy and the Labyrinths of the Mind*. New York: Oxford University Press, 2005.

List of Oppositions

List of Paradoxes

Index

FROM THE PUBLISHER

A name never says it all, but the word "Broadview" expresses a good deal of the philosophy behind our company. We are open to a broad range of academic approaches and political viewpoints. We pay attention to the broad impact book publishing and book printing has in the wider world; we began using recycled stock more than a decade ago, and for some years now we have used 100% recycled paper for most titles. Our publishing program is internationally oriented and broad-ranging. Our individual titles often appeal to a broad readership too; many are of interest as much to general readers as to academics and students.

Founded in 1985, Broadview remains a fully independent company owned by its shareholders—not an imprint or subsidiary of a larger multinational.

For the most accurate information on our books (including information on pricing, editions, and formats) please visit our website at www.broadviewpress.com. Our print books and ebooks are also available for sale on our site.

On the Broadview website we also offer several goods that are not books—among them the Broadview coffee mug, the Broadview beer stein (inscribed with a line from Geoffrey Chaucer's *Canterbury Tales*), the Broadview fridge magnets (your choice of philosophical or literary), and a range of T-shirts (made from combinations of hemp, bamboo, and/or high-quality pima cotton, with no child labor, sweatshop labor, or environmental degradation involved in their manufacture).

All these goods are available through the "merchandise" section of the Broadview website. When you buy Broadview goods you can support other goods too.

b

broadview press
www.broadviewpress.com

The interior of this book is printed on 100% recycled paper.